The Techno-Human Shell

A Jump in the Evolutionary Gap

Joseph Carvalko

The Techno-Human Shell
Copyright © 2012, by Joseph Carvalko.
Cover copyright © 2012 by Joseph Carvalko.

FIRST SUNBURY PRESS EDITION
Printed in the United States of America
December 2012

Trade paperback ISBN: 978-1-62006-165-7
Mobipocket format (Kindle) ISBN: 978-1-62006-166-4
ePub format (Nook) ISBN: 978-1-62006-167-1

Published by:
Sunbury Press
Mechanicsburg, PA
www.sunburypress.com

SUNBURY
P R E S S
Mechanicsburg, Pennsylvania   USA

ACKNOWLEDGMENTS

I would like to express my gratitude to those individuals who have helped make this book a success through their inspiration and suggestions, particularly to my wife, Susie, for her insights and patience for projects postponed; to my children, Cara Morris (a discerning reader/critic) and Joe Carvalko (cover designer); to Elizabeth Renfrow and Allyson Gard tireless editors of the lingua franca needed to bridge science, law and philosophy; to peer experts, Joel Marks, Professor Emeritus of Philosophy at the University of New Haven and a Bioethics Center Scholar at Yale University and Laura Pollander, Patent Agent; to the dedicated team at Sunbury Press for all their thoughtful expertise, including Lawrence Knorr, President and Publisher and Tammi Knorr, VP of Marketing & Author Relations; and to those individuals who inspired my life-long interest in science and engineering, Dr. Richard Kasper, Kendall Preston, Dr. Marcel J.E. Golay and Emil Bolsey and especially my parents, Joe and Lucille, who never stopped believing in the power of the creative mind.

CONTENTS

Sunlight[1]
by Jim Harrison

After days of darkness I didn't understand
a second of yellow sunlight
here and gone through a hole in clouds
as quickly as a flashbulb, an immense
memory of a moment of grace withdrawn.
It is said that we are here but seconds in cosmic
time, twelve and a half billion years,
but who is saying this and why?
In the Salt Lake City airport eight out of ten
were fiddling relentlessly with cell phones.
The world is too grand to reshape with babble.
Outside the hot sun beat down on clumsy metal
birds and an actual ten-million-year-old
crow flew by squawking in bemusement.
We're doubtless as old as our mothers, thousands
of generations waiting for the sunlight.

Now I've argued this is not genesis; this is building on three and a half billion years of evolution. And I've argued that we're about to perhaps create a new version of the Cambrian explosion, where there's massive new speciation based on this digital design.—Craig Venter [2]

PREFACE

If we study the great works of art, sculpture, and literature, we learn that our ancestors looked like us physically, thought like us, and perhaps felt much as we do. After all, we largely inherited their design: two arms, two legs, a body, and a head filled with matter adept at abstract thinking and feeling: a design which allows adaptation to the environment and survival into the next generation. However, as the human body tumbles headlong into the future, the fact that it and its ancestors may look familiar may be as far as resemblance goes. On the horizon of those now being born, their maturation, their death, and their children's natural cycles of birth, will be driven as much by technology as driven by the genes cast in the DNA backbone over 3.5 billion years ago.[3] Respected futurists have concluded that we have gone beyond our ability to retard this technological fate as we inescapably move toward a singularity—that event after which nothing looks like what came before. But is it really too late to change course? Must we resign ourselves to a future that will include successors that may look like us, but at their core will not represent the current model, nor represent what has heretofore passed as human? And if so, what steps should we take to push against devolving into a civilization of post-post-moderns that will look back upon their ancestors as Neanderthal cousins?

When the planet formed 4.6 billion years ago, oceans, mountains, and flat lands were created. Approximately 2.5 million years ago—relatively recently on the cosmic scale—*Homo genus* appeared, followed 2.3 million years later, *Homo sapiens*—those upright creatures we see in the mirror every day.[4] We have come to know ourselves not through mirrored reflections, but through conscious observation, thoughts, emotions, and self-awareness manifesting in a developed persona that exhibits a combination of openness, extraversion, neuroticism, agreeableness, and conscientiousness.[5] Combined, this persona learned to adapt to ever-changing circumstances to appreciate cause and effect, to invent tools, weapons, and skills and, over time, language that passed on the non-genetic essentials needed for its persistence in a sometimes hostile world. In the span of six to ten thousand years, language paved the way for philosophy, physics, chemistry, biology, medicine, computation, and the proliferation of know-how called technology. Over the course of the last five hundred years accumulated knowledge changed not only our understanding of the world outside ourselves, but changed our understanding of the world inside ourselves. [6]

The curves of scientific knowledge and technological progress bend increasingly skyward, every year more steeply as we reach for the stars from which we originated. Is our current track yet another indication of non-genetic essentials serving as tools, and language once did—this time transcending existential biological limitations, finally eschewing the *Homo sapiens* within the borders of our epidermis, to be reborn as altered creatures—as post biological *Homo sapiens*?

A central thesis of this book is: *As computers with ever increasing computational power of the famous Watson IBM computer spiral downward in size, the wholesale incorporation of these devices into the anatomy will become as common as a pill ingested, a vaccine injected or a body pierced.* [7] In the first phase, these will be installed first for medicinal therapy, then for diagnostics, followed by physical and mental enhancements. In the second phase they will be installed to more efficiently interface with a digital evolution.

It is in this second phase when Darwinian evolutionary rivers will merge with the rivers of intelligent designers, represented by scientists, programmers and engineers who will fuse organic natural biology, synthetic biology, and digital technology into a unified whole that future generations will deem their anatomy. The merger will serve to afford greater intelligence and, longer, healthier lives. In exchange, we will relinquish actual autonomy for apparent autonomy, where what was once considered "free will" will be supplanted by the deterministic logic of machinery somewhere in the mainstream of our unconscious.

Although in-the-body technology will have an explosive effect on commerce, entertainment, and employment, in the near term the concentration will be on medical devices, such as the innocuous pacemaker (essentially a working silicon-based computer, with sensors, memories, and a stimulation device with telecommunications to the outer world). In a second epoch, these devices will be gradually down-sized by advances in synthetic DNA, molecular- and nano-sized processors, each deployed alongside and within cells and organs as permanent non-organic, internal adjuncts to our anatomy for use as: nano-prosthetics, nano-stimulators/suppressors, artificial organ processors, metabolic and cognitive enhancers, and permanent diagnostic tools to ensure our physical and psychological well-being as we head toward a practically interminable lifetime. [8]

Will a wide-spread practice of installing technology into the body fundamentally change human essence? Our sense of self-sufficiency, authenticity, or individual identity? Will it change that numerical identity, the one "I" as some static aspect of ourselves (as self-consciousness as idealized by Locke)? Or will it change our narrative identity, our unseen internal human form, to eventually redefine what it means to be human?[9]

The digital revolution altered our social reality (just compare the habits of someone born a generation ago to someone born four generations ago). But today, most computers are at our fingertips. Darwin insisted "*natura non facit saltum*"—nature does not make leaps—but what happens when computers are moved inside our core, internalized, and become ubiquitous subterranean assistants function within our anatomical structure?[10] Will a new exterior reality emerge? Or will it go unnoticed, like the daily 80 milligram aspirin, acting undetected on our physiology? Social philosopher Francis Fukuyama suggests what Darwin might have concluded had he been alive today:

4

... while Aristotle believed in the eternity of the species (i.e., that what we have been labeling 'species-typical behavior' is somewhat unchanging), Darwin's theory maintains that this behavior changes in response to the organism's interaction with its environment. What is typical for a species at one particular moment of evolutionary time; what came before and what comes after will be different.[11]

This book will explore the question: Will the course of accelerating technological advance cause a subtle revolution in the human form—one that shifts the foundations of *Homo sapiens*, forming post-*Homo sapiens* who will have failed to record the journey, and thus making it impossible to return to a time, when humans roamed the Earth?

Cy·borg (sbôrg) n. A human who has certain physiological processes aided or controlled by mechanical or electronic devices.[12]

INTRODUCTION

According to Greek Mythology, the gods ordered Prometheus and his brother Epimetheus to create humankind. Epimetheus, being the more impetuous, took the lead, but rather than creating the species directly, formed animals first, to which he gave fur, feathers, wings, and shells, leaving his second creation, humans, naked. Prometheus came to the rescue, first helping this new creature stand erect among all beasts, and then he traveled to the sun, where he ignited a torch and bequeathed fire to Earth's new subjects so that camp fires would light the nights and enemies would be brought out of the darkness.

Fire changed humanity. It cast upon it light and shaped a being whose quintessence could not be separated from this new force—this new technology, which inevitably led the way out of inferiority—to stand tall against the forces of nature and other creatures. But it wasn't until the Industrial Revolution that the tools of technology began to inundate this new life. Some would say this was for the greater good, but it carried in its undercurrents unintended consequences. If modern technology were suddenly wrested from us, it is certain our days would be numbered and we would not survive more than a few days, weeks at most. Technology constantly shapes us as intellectual, moral, and social agents, yet when we first held fire in our hands, not even Prometheus could have predicted the extent to which we have evolved in this the Twenty-first Century.

The transition from human to posthuman will occur in two stages: the first memetic, where the current technological culture fades into a society, loosely in control, but largely accepting and dependent on in-the-body technology for its health, enhanced abilities, security, social mobility, and economic condition. The second stage will be temetic, where control is lost and technology, as an autonomous agent will have woven itself into the fabric of the human anatomy, demarking a physical transformation, where humans will be part biological and part technological. [13] Max More wrote:

> The transition from human to posthuman can be defined physically or memetically. Physically, we will have become posthuman only when we have made such fundamental and sweeping modifications to our inherited genetics, physiology, neurophysiology and neurochemistry, that we can no longer be usefully classified with *Homo Sapiens*. Memetically, we might expect posthumans to have a different motivational structure from humans, or at least the *ability* to make modifications if they choose. For example: transforming or controlling sexual orientation, intensity, and timing, or complete control over emotional responses through manipulation of neurochemistry. [14]

We already have entered More's phase, hardly acknowledging that our survival depends more and more in vital respects on technology, whether health (life-saving drugs, bone-marrow transplants, pacemakers, repairing genetic defects), financial security (banking systems, retirement investments), personal safety (local law enforcement, the vast military enterprise). The apparatchik of technology-driven institutions provides for medicines and medical treatments, food, refuge, and virtually every life sustaining good and service in the modern world, bar none. In every corner of the globe, individuals rapidly learn, use and contribute to technology, becoming fluent and dependent, much as our ancestors once became fluent and dependent on language or skilled in a trade. Future generations will increasingly adopt new programmatic restructurings as the technological world continues to change objects and sequences affecting: governance, work, play, health, and one's place in society.

Increasingly, governments, corporations, and universities operate, as if entitled by technological exceptionalism, to create forms of technology that go beyond all moderation, without consideration of conformity to cultural norms or humanitarian principles. We see this in the spread of genetically modified food, breeding genetically modified animals, and creating new life forms that serve no purpose other than to die from a deadly disease for medical research, chemical warfare, or scientific curiosity. Progress under a regimen of technological exceptionalism will move us into a world under the control of temetics or techno-memes, distant cousins to our cultural memes. In drawing the distinction between memes and temes, Blackmore says:

> While (human) brains were having an advantage from being able to copy--lighting fires, keeping fires going, new techniques of hunting, these kinds of things—inevitably they were also copying, putting feathers in their hair, or wearing strange clothes, or painting their faces... is there a difference between the memes that we copy—the words we speak to each other, the gestures we copy, the human things--and all these technological things around us?... Let's call them techno-memes or temes. [15]

The answer is "yes," there is a difference. Looking beyond transformations brought about by cultural memes, techno-memes will eventually overpower our successors, converting them into obedient hosts that carry out the techno-memes' survival imperative, not through adaptive natural evolution, but through adaptive replicative techno-modification. Blackmore's techno-memes will like a sleeping virus lie dormant in the recesses of our anatomy, until it acquires a critical informational content, a necessary and sufficient replicative design, and in the final stage parasitically bolts itself onto life-like biological machinery that had been implanted with good intentions to reduce illness, extend lives, and enhance the living experience. At that perilous moment, the transition from life controlled by cultural memes to life controlled by techno-memes will pass unnoticed, without as much as a whimper, where thereafter life will be lived according to a deterministic, algorithmic logic. In this stage, technological replication will surpass Universal Darwinism as the primary evolutionary paradigm for carrying successive generations into the future.

This book is not about fire or about the technology we have at our fingertips—the kind we reach for to send an email, make a call, pay a bill, or watch a movie. The technology addressed here will transform us, mimetically and temetically over the next century, when the forces of technology are applied to population control, commercial expediencies, medical therapeutics, and enhanced life. As we develop a threshold confidence in the utility of the technology that will produce these changes, we will begin to see such things as: RFID chips embedded beneath the skin for personal identification and for integrating ourselves into the consumer supply/demand chain; internal computer processors to countervail against geophysical changes (climate changes, genetically modified foods); prosthetics to replace malfunctioning, missing, or damaged body parts; carbon-based nano-prosthetics to seek and destroy diseases such as cancers; DNA bioengineered sequences to fix genomic defects; computer stimulators and suppressors for alleviating pain, depression or neurological diseases, such as Parkinson's disease; silicon-based artificial organ processors to replace any one of nearly two dozen major body organs; permanent analytical and diagnostic tools drawn from an array of bioinformatics technologies; sensors to keep vigilant over an organic anatomy that is under assault by deadly viruses, bacteria, and pollutants, from the time we begin to gestate until we die; specialized artificial intelligence processors to improve our integration into and our resilience in a technological world, by increasing our cognitive processes, such as the intellect; and finally processors to maintain the replicative superiority of the technology itself.

In the earlier stages of the evolutionary track, many of these technologies will be exploited, not only for their medical therapeutic value, but as enhancements to an already vigorous body, to improve the quality of lives, or to withstand the rigors of assault in the form of climate change, natural disasters or war. In other cases, the technologies will be exploited because we will be persuaded that it adds value to our ability to compete or to our social mobility—much like Facebook or MySpace does, or to facilitate commercial transactions, much like a credit card does—or really offers no apparent personal benefit, but is puffed up as indispensable to our well-being, by those who profit by lucrative monopolies the government crowns as patented inventions.

To be sure, other reasons may exist where individuals of their own accord or at the behest of organizations they work for, will have technologies installed into their anatomies. Today, implantable microchips for humans and animals (e.g., pets) the size of a grain of rice are interrogated by permanently located or hand-held readers. In the future, the size of the microchip will be along the lines of a white blood cell, 20 microns. The implanted technology today is largely RFID (radio frequency identification) that combines plastic, glass, and small amounts of metal. Human activity will drive the panoply of possibilities for in-the-body chip products. The fear of stuffing the body with chips that remain beyond the point of original utility will moderate when future microchips can dissolve after some useful implantation life or upon command—evaporate safely into the tissue beneath the skin. This is not science fiction. In 2004 it was reported that upwards of 160 employees working in the office of Mexico's attorney general, including federal prosecutors and investigators, had microchips implanted for obtaining access to secure areas in their headquarters.[16] In addition to unit security, currently these chips are marketed to help medical staff immediately connect to our medical records, to locate missing children or Alzheimer victims, and in combination with micro sensors, help monitor glucose levels in diabetics. [17] It is likely that in the near future, E-chips will automatically debit bank accounts, reward shoppers to stop at selected points of purchase, to verify we are not on the "no fly list", thereby speeding clearance at airports. The list is virtually unending.

Currently available non-electronic enhancements for the healthy include food supplements for energy or feelings of tranquility, growth hormones, drugs, cosmetic surgery, and things as benign as hair coloring. Products such as serotonin reuptake inhibitors (SSRIs) (e.g., Prozac, Paxil), and phosphodiesterase type 5 (PDE5) inhibitors (e.g., Viagra) might well be added as examples of drug-type enhancers. However, the near medical horizon includes bioengineered insertions into the cytoplasmic or nuclear structure of the cell or synthetic gene therapies that add not only a level of permanence to the human life qualities that food and drugs might improve upon, but in some cases go beyond to improving intelligence, providing more acute sensations (seeing, hearing, emotions), and find the Holy Grail of agelessness. Pharmacological progress will occur alongside the development of cybernetics—that is human-machine interfaces that will, by any measure, prolong, strengthen, lengthen, and boost human capacities —including lifespans.

The three drivers behind increasing computer implantation are bioengineers have reached a critical mass in converting computer technology into lifesaving and life enhancing products; computers and their software, the backbone for these products are becoming ever more computationally sophisticated based on the accumulation of scientific knowledge; between 2012 and 2020 processor speeds will increase from an already astounding 40 billion operations per second to 330 billion operations per second; and the substrates, that is the material part of the computer, now measured in micrometers or the size of blood cells, will dive deeply into the realm of wafer-thin graphene (carbon) and MoS2 (molybdenum disulfide) measured in nanometers, just a few atoms thick.[18]

Recent developments in decoding the human genome have led to the invention of synthetic DNA, and from such technology, the prospect of a prosthetic genome, which is hastening the day when life forms can be made entirely from non-living materials. These new technologies will transcend the inorganic chemistry of silicon and carbon and move into the realm of organic chemistry, eventually leading to the control of mechanical apparatuses and molecular processes via computational technologies, which will straddle the internal world, somewhere between that of

11

nature and human crafted human-machine artifacts. The macro-prosthetics of today (e.g., the pacemaker) will be reduced in size and improved upon by the nano-prosthetics of tomorrow, some manufactured in vivo (within the body), some pre-manufactured and embedded by physicians—in either case to live within our human shell. These quasi-natural artifacts coupled with developments in artificial intelligence bring the potential for smart nano-prosthetics to well within the horizon of the next half-generation.

Medical, legal, and techno-social commentators (e.g., Kurzweil, Joy, Vinge, and others) continue to explore the ramifications of the unprecedented magnitude and rate of change of bioengineered technology on individuals, society, economics, and law related to organ transplants, drugs, DNA therapies, or stem cells—technologies that are largely biological or nature-based. For the one-hundred years preceding World War II, surgical technology, medical instruments (such as the x-ray) and materials (such as the suture) exemplified the level of medical advancement.[19] Following WWII the advancement of medical arts accelerated, where one discovery or invention—some grounded in biology others in physics—compounded progress each year. In the background, advances in pharmacology, material sciences, bioengineering and most notably computer technology were applied to diagnosis, therapy and prosthetics. For the next forty years, from about 1950 to 1990 the rate of progress was unprecedented, improving diagnosis, improving the quality of life and longevity throughout the world. For evidence of this last point, we need not read medical or science journals, but simply look around at the availability of used organs—from hearts to kidneys, the ubiquity of dental implants, the prosthetics for those who have lost limbs in accidents or wars. In the future the incorporation of in-the-body technology will be driven as much by the fact that people will live extraordinarily long lives, in virtue of advances in biochemistry and bioengineering, but also resulting from the body's dependency on the technologies of computer engineering, communications, and nano-based materials. Let us look at two examples of the types of technologies currently available to underscore the changing state of medical technology over the last several decades.

Over the course of many years, the nation had been made aware of Vice President Dick Cheney's health problems and remedies, the latter charting 45 years of techno-medical progress in heart assisted devices.[20] Cheney had his first heart attack in 1978, at age 37. These were followed by attacks in 1984, 1988, 2000, and 2010. After the 1988 attack he underwent a four-vessel coronary artery bypass graft, then in 2000 coronary artery stenting, followed in 2001 by both a coronary balloon angioplasty and an implantable cardioverter-defibrillator.[21] In 2005, Cheney underwent a repair of a popliteal artery aneurysm, and then in 2010 he was provided with a left-ventricular assist device to pump blood throughout the body.[22] In an interview at the time, Dr. Kirk Garratt, clinical director of interventional cardiovascular research at Lenox Hill Hospital, told a reporter at the New York Daily News, "He really doesn't have a pulse, but he has blood pressure because blood is being pumped out from his ventricle into the aorta at a constant pressure." [23] Cheney was without a pulse for fourteen months, until he finally underwent a seven-hour heart transplant procedure. Remarkably he was seventy-one years old. Michael Anissimov of the Singularity Institute told me:

> The application of cybernetic prostheses to the human body changes our notion of what "human" means right away... Already there are concerns over what might happen when runners with prosthetic legs begin to outrun those with flesh legs. Already there is wonder over whether certain Olympic athletes may have been genetically modified. Though no one would claim that these people are post-humans, they are arguably trans-humans, as they are being given abilities through technology that no human has ever been given before, which would deeply surprise people of only a few generations ago.

Case in point: Olympian Oscar Pistorius's lower legs were amputated below the knee before he turned one year old, but that did not stop him from competing in the 400-meter run in the 2012 Olympics Games, using artificial legs made from carbon graphite, which, spring-like, allowed him to run and even jump.[24] As the entire world watched his stunning one-lap effort, it proved to be not only a remarkable example of the state of prosthetic technology, but a recognition of a legitimate adjunct to the anatomy in a sports culture that typically shuns any patently artificial competitive advantage. Pistorius told one reporter, "I didn't grow up thinking I had a disability, I grew up thinking I had different shoes." And as much as he expressed the willing acceptance of technology into his own anatomy, it reinforced the notion of artificiality as a new norm.

Cheney's and Pistorius's experiences are the outcomes of Nobel Prize winning breakthroughs that have fused physics, information theory, biology, and communications. These innovations feed directly into impressive medical applications that have to do with organ rehabilitation or repair or solving complex bioengineering problems that come with using new materials at the macro-organ or micro-cellular level. The marriage of computers, nanotechnology, and genomics increasingly merge into bio-machinery that operates under artificial intelligence paradigms of the sort that allow for deciphering the human genome and the promise of everything from genetically tailored pharmaceuticals to genetic reorganization and repair.

In addition to the obvious benefits to the medical sciences, technological advances of the Cheney-Pistorius type generally spark a wide range of social, legal, political, and cultural issues. Pistorious's first international competition in 2007 immediately drew claims of an unfair advantage over able-bodied runners. That year, the International Association of Athletics Federations (claiming that Pistorious was not the motivating factor) banned the use of "any technical device that incorporates springs, wheels or any other element that provides a user with an advantage over another athlete not using such a device."

The new rule denied Pistorious from competing in the 2008 Olympics. He took his case before the Court of Arbitration for Sport, which eventually ruled that no evidence existed that there was any net advantage over able-bodied athletes, paving the way for his 2012 competition. We might think that Cheney and Pistorius are the outliers in society, and certainly their stories are dramatic, but many of us already incorporate products that combine computers, nanotechnology, pharmaceutical and biological products into our anatomical shells either on a permanent or occasional basis; in effect, some of us are already enhanced.

Medical disability technologies diffuse into society through those who need them, such as pacemakers for victims of heart block or arrhythmias, artificial limbs for amputees, and cochlear implants for the deaf. Technologies that enhance human performance, such as Pistorius's prosthetic, have their original impetus in supplying a solution to a medical disability or in providing a military advantage, which will we speak of in a later chapter. But technologies that enhance otherwise normal performance will inevitably, under elective circumstances, be employed by those who can afford the device. Questions will be raised as to whom and under what conditions should anyone have the right to upgrade their cognitive or physical attributes.

In futurist circles comprised of transhumanists and technoprogressives, the acronym GRIN stands for genetic, robotic, information, and nanotechnology.[25] The government has coined the acronym, *Nano-Bio-Info-Cogno* (NBIC) as shorthand for the convergence of nanotechnology, bioengineering, information systems, and cognitive technology.[26] Both these lists imply the emergence of nanotechnology, bioengineering, and information technologies in the forms of computers, artificial intelligence, and communications to enhance human performance. In these fast-moving disciplines we find developments in synthetic DNA for attacking disease and providing therapies, nano-sized miniaturization of computers for diagnosing illness, communicating out of the body, delivering drugs, or special killer molecules directly to the cell, artificial intelligence and software breakthroughs that regulate the numerous bodily processes. These technologies are applied to increase the quality of life for disease victims and they increase their life expectancy.

As indicated, we have already seen that through implanted devices victims of accidents, birth defects or diseases can return to a level of performance or sustainability that keeps them going. There are also nootropic (smart) drugs, neural prosthetics, and physical prosthetics, being developed to enhance performance—for example, for soldiers looking for an edge in combat. Many of the NBIC technologies for enhanced performance will be applied to military personnel. Noah Shachtman, who writes Army and Marines, Gadgets and Gear for *Wired Magazine Online*, says: "'Neural prosthetics' and 'smart drugs' will make them battlefield geniuses."[27] In a report from the U.S. Army Natick, Research, Development, and Engineering Center entitled: *Future Soldier 2030 Initiative*, U.S. Army General George Casey writes:

> The goal of our Army is to continue the transformational process of building a campaign quality expeditionary Army that can support our combatant commanders in the challenges of the 21st Century across the full spectrum of conflict... Two possible areas (as regards wound/bleeding management) are available. First area is wound treatment and the second is the use of autonomous & self-injected drug delivery systems. The wound treatment would incorporate means to self clean open wounds and seal the wound sites. *The second area would potentially use implantable subcutaneous auto-injector with rapid reconstitution packages to treat various types of chemical, biological, or other threats through wearable and implantable MEMS-based devices.* The subcutaneous systems could also be incorporated into small externally worn devices.

In civilian and military applications advanced communications systems will be employed, not unlike the Internet, to shuttle data to and from vast databases, located not only in cyberspace, but also in an anatomical space referred to as the "human shell", creating, as transhumanists would predict, cyborgs. [28]

No more gods, no more faith, no more timid holding back. Let us blast out of our old forms, our ignorance, our weakness, and our mortality. The future belongs to posthumanity.—Max More, *On Becoming Posthuman* [29]

SNAPSHOT OF THE FUTURE

What if one hundred years ago someone inserted a specially designed mechanical clock into a man's chest and brought him back to life? What would our ancestors have called the survivor? A robot, Frankenstein's monster? [30] Today, no one thinks twice of anyone dying and being resurrected after an emergency installation of a pacemaker. Biomedical devices sustain and extend hundreds of lives, reduce pain, help the physically challenged experience nearly normal lives. But we stand at a technological precipice where computers, with the computational power of the famous IBM Watson computer, will have direct access to our neurological anatomical circuits and change our moods, provide sight to the blind, and supply sound to the deaf. In the slightly longer term, as computer technology advances, the benefits will not only remediate a dysfunctional organ, but improve upon our biological framework, helping humans avoid the end-of-life vagaries of a probabilistic Darwinian gene set, in favor of a programmed *Ex Machina* order that extend the current normal life span, many, many decades. The wholesale incorporation of these devices may someday become as common as a modern-day vaccine and may pose an existential threat to natural biological evolution. At what point does the idea of an actual cyborg-assisted-life or a life-assisted-cyborg change our attitude or opinion about what the notion of "human" means? What might the social and legal implications be?

Advances in future medicine and medical technology are harvested from the fields of bioengineering, computer engineering, and materials. The motivation for innovation stems from a desire to reduce human suffering, to increase life spans, and to enhance the lives we lead. The secondary motivation comes from the scientist's quest to discover and the technologist's to invent, as a way of self-actualizing their professional lives, careers, their fame, and fortunes. Next, institutions, such as universities, corporations, and government research operations seek to fulfill their missions, survive economically, and compete with others to advance science and technology. Finally, commercial enterprises such as manufacturers of devices (e.g., RFID chips and pacemakers), purveyors of goods and services (hospitals, pharmacies, and insurance companies), look for ways to improve upon their business models, and to make them more efficient, by reducing the costs associated with hospital admissions, drug delivery and on-site medical examinations. We experience this when we have to struggle through a phone prompts, self-checkouts at the local retailer, or paying bills online. Each of these stakeholders will seek out opportunities in the future.

Let us conjecture what life might be like a decade from now by visiting with a young girl named Eve.

It is late August, and Eve just turned eighteen. Two months ago she graduated high school and now looks forward to starting college. She has a lot to do, not the least of which is to buy clothes for the new school year. She figures that she will use her grandmother's generous graduation gift, enough money deposited in her bank account to keep the freshman in fashion for an entire year. She picks up a picture of her grandmother on the dresser and thinks about how little Grandma has changed since the picture was taken over thirty years ago. She could only hope to look as good in her eighties as her grandmother does. Her grandfather, Dr. Robert Morgan, an early adopter of new-fangled technology, left her a keepsake to start college. It's a new smart chip that will help download books hassle free, but more important to Eve, it will help her spend the money that Grandma deposited into Eve's account. She begins to read the instructions for her latest addition to an arsenal of communications APPs and devices she owns.

"Smart Chip is the free web tool from Millennium Express that will provide you with a secure and convenient way to manage your shopping experience. You will never have to type another URL, User ID, password, or shipping address again! Here's how it works: Store your personal profile, credit card information, and favorite URLs directly onto your Smart Chip.

The Smart Chip from Millennium Express is an intelligent RFID computer microchip that holds a certificate of authenticity. It provides you with additional security and personal file access by you and those you authorize, whether at a local or distant merchant, online, at airports or hospitals, for all cards issued after 1/1/2032. Simply follow the instructions below, and in no time the chip will be securely implanted to provide you the freedom to shop without worrying about paying or passing through the checkout counters again.

What is the Smart Card Reader?

A Smart Card Reader is installed in the retail outlets of more than 500,000 Millennium subscribers, airport terminals, and hospitals throughout the world, allowing you to be automatically identified as a Chipmember and qualified for services such as: shopping at your convenience, day or night; and removing from retail displays whatever merchandise you wish to purchase; and leaving the store without having to pass through its checkout counter. If you choose to shop online a Smart Card reader, connected to your computer through the USB port, allows you to take advantage of a higher level of security and allows you to utilize the Smart Chip as your main access to your favorite online store without the hassle of checking out. Just point and click and the product will be shipped to you overnight and your account will be automatically debited.

If you sign up for our Medi-Alert program, your vital medical and personal information will be collected throughout the world and placed on our database, making it accessible as you enter any subscribing hospital or emergency room throughout North America and most of Europe and Asia.

How does Smart Chip work?

To take advantage of the Smart Chip, remove the sterile pod from the package labeled 'Smart Chip Pod' as shown in the instructions on the back of the package. As shown in figure 1, put the POD between your right thumb and index finger and place the POD on your left bicep. With a slight pressure, press the POD into the bicep. The specially loaded POD will painlessly and automatically penetrate your skin. Open the package labeled 'Alcohol Pad', remove the pad, and apply for 30 seconds to the area where the Smart Chip has been installed. Your Smart Chip ID is pre-loaded on the chip and when inserted you must sit near a computer or your smart phone—either of which will pre-load your Smart Chip into our system and activate your subscription. Once the application is on our Smart Chip Server, you may access 'My Profile,' 'My Cards,' 'My Favorites,' and 'My Notes' directly from any device.

Now you are ready to let Smart Chip help you surf the Web, shop in stores or online, download pay-only sites on demand, such as movies or music, and manage your finances with ease, security, and convenience! As you access various websites, the Smart Chip will memorize your favorite URLs, log-ins and personal data so you'll never have to input them again!"

We have ventured into the future to get a glimpse of Eve's new found freedom from having to deal with the inconveniences of passing through checkout lines—but virtually all the technology mentioned to achieve this newfound convenience is already here. For example, new credit cards have RFID technology so that card holders no longer have to swipe their cards; passports now have RFID to identify the passport holder; and now driver's licenses will employ the technology allow for effortless electronic toll-collection. Perhaps the most stunning application was reported in 2004, where a nightclub in Barcelona offered VIPs a microchip for implantation under the skin, which when scanned would guarantee entry and provide access to a debit account for purchasing drinks. [31] Soon the idea popped up in its club in Rotterdam, Holland. [32] The chips are routinely installed by doctors using a needle to insert the chip a few millimeters beneath the skin. The procedure takes only a few seconds.

Eve's application for the general consumer needs to await a critical mass of retailers and consumers to "buy into" this merchandizing paradigm shift. As we explore the technological possibilities that will greet us in the future, let us keep in mind that the acceptance of different technologies by public is uneven—some technologies will be available, but will lay dormant awaiting our collective willingness to use them. This has always been true about invention. For example, cameras took a generation and telephones took more than two generations to gain footholds in society.

In cases of a truly novel idea—one not yet introduced into the market—inventors can only conjecture who might first adopt the novelty. Famously, Thomas Edison filed a patent application for a machine that "would do for the eye what the phonograph does for the ear." [33] But an employee, William Dickson, reduced the idea to practice in a device that became known as the Kinetoscope. After an extended trip abroad, Edison returned to his labs where Dickson presented him with a short talking picture. Edison was impressed by what he called the "Wonderbox." The two of them refined the product over the next several years, but they disagreed about the market. Dickson pushed for movie projection on the big screen, while Edison decided that peep shows were the better outlet, which he pursued. This delayed the availability of big screen motion pictures, and illustrates a common shortcoming of inventors—often, they do not how their inventions will eventually be put to use.

In 1962 Everett Rogers, studied many hundreds of cases related to the adoption of technologies—mostly concerning farmers and the medical community—and how, under what conditions, and the rate at which his subjects adopted new ideas, practices, or objects. [34] He advanced a theory that the diffusion of technology into a society relates to individual's knowledge of the innovation, the force of persuasion concerning the beneficial aspects of the innovation, his or her decision to accept or reject the innovation, using the innovation, and confirmation—in which the innovation takes a relatively stable position in his or her life. For the kinds of products we are discussing here, the adoptive conditions and the adoption rate will take place within well-defined social situations and structures, in which people will have little autonomy or choice in eschewing the innovation, for instance military application, medical intervention, and relentless market persuasion. This last force is well understood by those under the unremitting battering of the pharmaceutical industry's two billion dollar television advertising budgets.

The embedded computational and communications technology that Eve employs will develop parallel to efforts genomic technology, which will utilize gene sequencing to write personalized genomes as a means of developing cancer drugs tailored to a specific individual. The use of second generation sequencing platforms have kept prices too high for personalized sequencing, but *Next Generation Sequencing* (NGS) technology, has driven costs down to between $5,000 to $10,000 (as of 2010), and it promises to further reduce costs to within range of $1,000. [35] NGS technologies will cost-effectively sequence complete genomes, distinguishing genetic fingerprints to determine the etiology and the behavior of "personalized cancers." Initially the technology will run in conventional laboratories, but soon these will be tested in clinical laboratories, cutting costs dramatically. And the expectation is that future generations will employ in-the-body laboratories, current micro-labs on the order of one centimeter square (the size of a pinky fingernail), which will make the current model of the pathology lab all but obsolete. [36]

These large-scale in-the-body biological laboratory initiatives will not halt replacement therapies, such as organ replacement or the manufacture of cells via stem cells for growing new tissue, nerve cells, and organs. And it will not slow down developments in computational and communications technology, which will serve in capacities where pharmaceuticals, genomic biochemistry, and bioengineering cannot—for example instantly transmitting diagnostic findings as soon as the biological clues are deciphered.

Computers and communications have the capacity to augment diagnosis and therapy by capturing data about how the body is responding to a therapy and telecommunicating the result to a central data base for analysis and observation by medical professionals. There may be considerable advantages in measuring the well-being of major organs, such as the heart, the kidney, pancreas or liver, and transmit the results to a remote server. Although drugs or therapies keep us healthy or from falling into states of un-wellness, there may not be any reasonable way to insure that we maintain stasis, but for devices that operate as personal medical laboratories, in real time, taking stock of the analytics: the production of enzymes, whether we are fighting infection or suffering from too much or too little medication.[37]

Dr. M.C. Roco, Chair of the U.S. National Science and Technology Council Subcommittee on Nanoscale Science, Engineering, and Technology writes:

Integration of NBIC tools is expected to lead to fundamentally new products and services, such as entirely new categories of materials, devices, and systems for use in manufacturing, construction, transportation, medicine, emerging technologies and scientific research. Fundamental research will be at the confluence of physics, chemistry, biology, mathematics, and engineering. Nanotechnology, biotechnology and information technology will play an essential role in their research, design, and production. Industries increasingly will use biological processes in manufacturing. Examples are pharmaceutical genomics, neuromorphic technology, regenerative medicine, biochips with complex functions, molecular systems with multiscale architectures, electronic devices with three-dimensional, hierarchical architectures, software for realistic multiphenomena and multiscale simulations, processes, and systems from the basic principles at the nanoscale, new flight vehicles, and quantitative studies with large databases in social sciences. Cognitive sciences will provide better ways to design and use the new manufacturing processes, products, and services, as well as leading to new kinds of organizations.[38]

The human anatomy contains 50 major organs, each of which will be affected in some way by transforming technologies—analyzed, diagnosed, prescribed treatment. Until most of this process is self-contained in the human body for purposes of automatically administering therapy, there will be an extensive communication from in-the-body computers to the outer world, via telemetry, over the Internet or proprietary channels, where larger computer systems will perform complicated analysis and recordkeeping. The external world of communications will become an adjunct to our anatomy, ensuring our individual, as well as our collective, well-being. Large amounts of data, collected from millions and eventually billions of people will not only ensure a single person's health, but the health of entire populations. It is in this stage of public health that medicine will be completely predictive, preventive, personalized, and participatory.

Just as Eve's smart chip evaluates data from a host of outer-body commercial sources, other in-the-body digitized information technology will acquire and evaluate data from our cells to create a personalized internal medicine cabinet. In some instances the data will be consolidated and contextualized. We can imagine that a future body in need of a prescription drug might forwarded that information to a pharmacy, which then determines if a doctor's authorization exists and if so, fill the prescription, and if not notify the doctor.

But what happens when the evaluated data reveals a life threatening disease or one that may potentially be contagious? The information will be sent to our physicians, —at least in the near term—but depending on the level of seriousness (for example a condition that can immediately affect a larger population) it is likely it will be sent directly to an agency such as the National Center for Disease Control.

What price will we pay in individual liberty and privacy for this level of social utility? More broad effects would be to alert public health officials of impending pandemics, but more personal ramifications might affect the potential transmission of viruses like the common cold from mother to child, or "private" diseases such as AIDS or STDs? These issues of privacy must be considered. Each of us will be connected, not only in a social sense, but through the networks that will read, in some cases analyze, diagnose, and prescribe for our so-called well-being, and the well-being of society at large. This raises privacy and security issues that make present day concerns seem benign in comparison. And while drug companies might look askance at personalized drug therapies due to their current high development costs and low payback, public health authorities and insurance companies would welcome systems that allow them to peak in to one's internal anatomy and physiology on an ongoing basis.

What happens when the technology of computation, process control and management, drug delivery, and communications embeds itself in the deep recesses of our biological apparatus, subtly keeping us sharp, vibrant, alive, and well? We would certainly undergo a revolution in social reality. One might see the thrust here as the evolving Million Dollar Man— the character born of technology—except assumptions here are based in fact and the sound predictions from logical extensions of what we can reasonably forecast about the future of life-dependent, in-the-body biomedical technology.

Unlike stock markets or horse races, technological advancement can be reasonably predicted (e.g., Moore's law that microprocessor semiconductor density would halve every two years). When semiconductors were invented in the '50s and became prevalent in the '60s, companies started developing the next generation—the microprocessor—nearly a decade before they emerged in handheld calculators. It was a few short years before these processors invaded practically every home appliance, car, plane, personal computer, phone, pacemaker, prosthetic limb. Scientists predicted when processors would reach particular process speeds, attain a certain physical size or reach a memory capacity. Less predictable were the inventions that made such items as the personal computer easy to use: the mouse, browsers, and search engines. The future holds an astonishing array of products, some extensions of what we have today and that are accurately predictable and new products as yet imagined.

Humans will find themselves deeply integrated into systems of machines, but they will remain biological—at least for the foreseeable future. This does not mean that machinery (broadly defined to include the software, again broadly defined) will not invade our organs to as deep a level as the nucleus of a cell, its chromosomes, and resident DNA. So what does the world look like twenty, forty and sixty years from now, when changes in technological advances compel a healthy population to incorporate biomedical computational devices to manage their physiological status? Let us conjecture based on where current technology is headed, and based on a sampling of a few noted authorities as mention throughout this book:

2015-2025—Most individuals in developed nations will have one or more aspects of their health directly connected to combination technology. These might include devices that use home computers to measure and record vital signs, such as pulse rate, blood oxygen levels, blood pressure and blood glucose.[39] A significant number of individuals in developed nations will have some form of embedded device, such as an RFID chip or bio-processor. These devices may be in combination with: (a) pharmacologicals, for managing and controlling some element of health, (b) patient identification, including a summary of a patient's records either also embedded or automatically accessible when in proximity of an interrogating computer, (c) maintaining patient stasis or stability, (d) diagnosing conditions and reporting on adverse conditions via the Internet or private network, (e) alerting medical authorities to dangerous drops in vital signs, heart rate, heart electrical activity, hormones, enzymes, or changes in the immediate environment (potential airborne hazards, e.g., chemical pollutants). Medical devices will produce and dispense electrical excitations, drugs, enzymes and hormones—insulin for example—from embedded in-the-body pharmaceutical factories. Most individuals in developed countries will have a full awareness of cognitive and biological technological forces (akin to today's familiarity with the Internet, laptops or iPads) that are operating their lives (internal and external), and enabling healthier lives, longevity, creativity, and decision-making.

2025-2040—Reactive medicine of today will be replaced by what has been popularly coined "P4" medicine: predictive, preventive, personalized, and participatory. Most medical examinations will be performed in the convenience of one's home, via large wall screens and Skype-like communication's hook-up to medical devices, some remote and others either internal to the body or in the home itself. Looking at the slope of the progression of the key technologies required to advance to the early stages of P4 medicine suggests that it will occur after 2035. At least part of the progress comes from innovative sensors that will be installed in the bathroom to detect the presence of cancers and other diseases before they even begin to manifest on a level that calls attention that something is wrong.[40] Other developments in genetic based polymers and nano-carbon materials will interface within the human organism, detectable through computer-like processors, operating in the nano-sized spectrum, running biomedical diagnostic and therapeutic algorithms to determine the presence of disease or the desired mix of personalized drugs for a course of treatment.[41] Application of any drug, today, is not localized and rather than being targeted to an intended site, often invades multiple organs. New devices, largely nano-electromechanical drug delivery systems coupled with synthetic DNA will solve this problem and revolutionize medical prescription delivery.

In the non-medical arena, almost everyone in modern society will court some type of smart RFID technology that identifies them by name and address and reports things as mundane as shoe size. More significantly, there will be "hook-ups" to banks, retailers, airline check-ins, voting booths, access to public buildings, and immigration.

2040-2060—The human body will be significantly infused with analytical labs-on-a chip for warning of disease in hours, rather than learning about an illness months or years later, allowing the individual to live more durably, resistant to stress, and the effects of aging. In-the-body technologies will compensate or eradicate most physical and mental challenges. Scientists will perfect downloading the brain into a memory substrate.

2060-2090—Supercomputers the size of a human blood cell will be incorporated into every human body, promoting cognitive function and enhanced intelligence, and stabilizing overall well-being and preventing disease. Interfaces between the human brain and machines will transform the relationship between humans and machines. By merely conveying one's thoughts one will be able to directly communicate with computer application, such as a word processor or a spread sheet, control machinery, enable entertainment devices, and play synthesized musical instruments without laying a finger on a keyboard of any sort.

2090-2130—To communicate using a form of human telepathy employing quantum computers; and if one could afford the prosthetic, to see three hundred and sixty degrees or watch video directly through a cortical implant.[42] Disease as we know it will be virtually eradicated, and there is a marked evolution in lifespans approaching several centuries.

It is not so startling that the future will witness near superhuman artificial intelligence in the world outside of us (we see that already in the Mar's Curiosity rover or the IBM Watson computer), but it is startling that through the integration of computational technology and human physiology, we stand to gain superhuman artificial intelligence in the world within. Drug and gene therapy will likely improve human performance and intelligence and will continue to increase immunity to a panoply of deadly diseases, but the eventual anatomically widespread deployment of the technologies of sensing, computation, and communication will change our species into a cyborg-like human, where practically everything we do will be guided in some way by the technology from within.

Expert medical computer systems already provide qualitative medical solutions and answers to medical diagnoses, especially those that depend on complex arrays or enormous quantities of data. In the future individuals will have a full complement of their own in-the-body computer sets to monitor health and determine whether he or she is tending toward a medical condition that puts him or her at risk. A part of the complement will consist of the expert system, which will serve as a peripheral to an in-the-body central communications server (CCS) that coordinates with a remote out-of-body medical server and emergency services. The CCS would also alert a dangerous condition to the individual or will take immediate remedial actions to mitigate a life threatening condition.

Expert computer systems will have the facility to obtain embedded sensor's data, retrieve relevant information from internal and external databases, and then interpret and draw inferences about medical conditions using decision trees, neural networks, and statistical inference engines. The general architecture of an expert system involves a problem-dependent set of data declarations called a knowledge base, and a problem independent program, which is called the inference engine. The data collected from sensors or other quantifications of anatomical conditions (e.g., temperature, electrical activity, enzyme, glucose levels), provide the information from which a likely condition may be determined or predicted. The inputs to the expert system come from organ-specific distributed processors that actively monitor "current status", through real time polling of embedded sensors. Pattern recognition systems, such as utilized in the form of neural networks and inference engines, provide for a high degree of accurate decision making, on a par with, if not better than diagnoses and prognoses achievable by a competent specialized medical professionals.

We once used our legs or horses to travel, only spoke with people face-to-face, and listened to people within shouting distance. At the turn of the twentieth century we first breathed in technology that would change the way we had lived for the better part of a few hundred thousand years. Individuals had relatively short lives limited to walking the earth or sailing the seas. The closest comparison between now and the future is when society changed from the horse-less carriage to the automobile, from face-to-face communication to the telephone, from ground transportation to the airplane, from the world before television and radio to one that has birthed the Internet, where we are constantly entertained and in touch with relatives, friends, and business associates. Indeed the changes going forward will be subtle, but nevertheless will shift the social paradigm for all future generations to such a degree that any reference to one's historical predecessors will be like a present-day student reading about god-like characters from ancient Greece.

Gatsby believed in the green light, the orgiastic future that year by year recedes before us. It eluded us then, but that's no matter-- tomorrow we will run faster, stretch out our arms farther...—F. Scott Fitzgerald, *The Great Gatsby* (1925)

MANUFACTURING NEW LIFE

The modern story of forming new life from whole cloth begins in the early 1940s, when, armed with the knowledge that there were in essence only two types of biological replicators— chromosomes and viruses, both nucleoproteins—Avery, MacLeod and McCarty mixed a dead pneumococcus bacterium, we will call this D, with a live bacteria, we will call this L, and found that it replicated into the dead bacteria D. This led to conjecture that the dead bacteria transferred basic information about itself to the live bacteria. Genetic material represents information, and like the information stored between the covers of a book, it is stored in a particular medium, is copied, read, and communicated. The team concluded that the biological material, DNA (observed by Fredrick Griffiths sixteen years earlier), was the same hereditary material involved in transforming the bacteria they had observed.

After the second World War, advances in electron microscopy allowed increasingly smaller projections of molecular structures. The structure of solids observed at the atomic level was for some time known to be organized into regular geometric patterns. When DNA was first observed through the electron microscope it displayed a regular pattern observed in other fixed atomic and molecular structures. James D. Watson and Francis C.H. Crick studied these projections and in 1953 reported, "We wish to put forward a radically different structure for the salt of deoxyribose nucleic acid. This structure has two helical chains each coiled round the same axis..."[43] This double stranded molecule was of course the DNA as it has come to be characterized and understood.

In the same year Watson and Crick made their historic discovery, a young Stanley Miller, one of Nobel Laureate Harold Urey's graduate students at the University of

Chicago, tested his professor's theory that conditions on primitive Earth favored chemical reactions that synthesized organic compounds from inorganic precursors. Simulating what he thought were those conditions, Miller passed an electric current through a flask filled with water, hydrogen, methane, and ammonia. Within the week he discovered that the experiment produced 20 amino acids—the kind found in DNA—demonstrating the ease with which the essential ingredients found in all life can be created.[44]

Twenty years after the Miller-Urey and Watson-Crick discoveries, a microbiologist named Ananda Chakrabarty manipulated two distinct life forms, bringing their genetic components together, and they manufactured a new bacterium capable of breaking down the components of crude oil from the genus *Pseudomonas*. It contained at least two stable energy-generating plasmids, where each plasmid essentially provided a separate hydrocarbon-degrading pathway. Because of this property, possessed incidentally by no naturally-occurring bacteria, Chakrabarty filed for a patent claiming bacterium that devoured oil spills. The patent office rejected the application based on the long standing prohibition against patenting forms of nature.[45] In 1980, the Supreme Court, in a split 5-4 decision, held that regardless of the wisdom of patenting life forms, the process by which these bacteria were created (human ingenuity) falls within patentable subject matter as intended when Congress drafted the most recent Patent Act of 1952.[46]

The Supreme Court went out of its way to avoid the long standing prohibition against patenting naturally-occurring physical phenomena.[47] From Chakrabarty forward, the relevant distinction in bioengineering patents would not be between animals, minerals, or vegetables, but between "products of nature", whether living or not, and the human ingenuity that led to that watchword "invention." Since that litigious upheaval, hundreds of thousands of genes, gene fragments, and the generation of interspecific mammalian "chimeras" have been the subject of patents. These later species developed from mammalian embryos and full term animals originating from embryo cell mixtures, chimeras began with rodents and rabbits in the early '70s, to reports of viable "geeps" (goat-sheep chimeras) in 1984. Throughout the 1980s, the patent office granted patents for gene therapies, transgenic animals, expressed sequence tags, antisense oligonucleotides and single nucleotide polymorphisms. [48]

In 1987, the patent office ruled in *ex Parte Allen* that under Federal law governing patentable subject matter, oysters that have been artificially treated to alter the number of chromosomes are patentable.[49] Of course this development was significant inasmuch as Allen's research focused on adding chromosomes to the basic architecture of a living species. We will revisit this development regarding the likelihood that same kind of technology might be used to add a chromosome to the human complement of forty-six.

In 1988, following the Allen decision, Harvard scientists Philip Leder and Timothy Stewart were granted a patent on a mouse that had been genetically altered to increase its susceptibility to certain forms of cancer. Referred to as a transgenic non-human mammal and named the "Nocuous" it turned a sharp corner that would open up new ground in the world of patenting genetically modified animals.[50] Never before had a whole, living animal been the permitted subject of a patent. But there would be plenty of other creatures created and patented, however. Until the early twenty-first century the world had only known the kind of species produced through *naturally occurring* DNA. [51,52] But that too will change—especially as we venture into the realm of *synthetic* DNA, which promises to produce life-like specimens engineered by man.

Synthetic DNA use chemicals to construct molecules that behave in DNA-like ways. In turn these are applied to non-natural utilitarian functions, such as medical therapy and biofuels. Recently, J. Craig Venter Institute assembled, modified, and implanted a synthesized genome into a DNA-free bacterial shell to make a self-replicating Mycoplasma mycoides.[53] There is no natural counterpart to this. Early applications will be in biofuels. Synthetic biology combines biology and engineering in ways harkening back to the early semiconductor days where silicon doped with impurities provided the means for controlling electron flow. Once semiconductors were arrayed in large scale circuits or computer processors, only then did software engineers come on the scene to digitize information technology in the twentieth century. The synthetic biologist today is at the stage of early "doping" and designing the networks and processes not found naturally, and in many respects, adding another sense to the conception of "computer science." Soon a new breed of "software engineer" will arrive to digitize information technology on newly created synthetic DNA platforms.

The most revolutionary medical advances during the latter half of the twentieth century have been brought about through technology, followed closely of course by medical practice itself.[54] The twenty-first century will be dominated by technology exploiting materials at the molecular and nano scale, many of which are not biologically based. But they will be combined or fused with synthetic DNA, which is biologically based, and together the confluence will do for medicine and the anatomy what steel and plastics did for the modern industrial complex, as it creates artificial organs, computer-like half-living cells and invented genomes.

Mark Bedau of Reed College, Oregon, says, "A prosthetic genome hastens the day when life forms can be made entirely from non-living materials. As such, it will revitalize perennial questions about the significance of life —what it is, why it is important and what role humans should have in its future."[55]

In March 2008, Craig Venter impressed a crowd assembled at a TED forum describing his work in the creation of a living bacteria. [56] Notice how he frames language in terms of digital technology:

... we've been digitizing biology, and now we're trying to go from that digital code into a new phase of biology with designing and synthesizing life... Our ability to write the genetic code has been moving pretty slowly, but has been increasing, and our latest point would put it on, now, an exponential curve. We started this over 15 years ago... And this was our first attempt, starting with the digital information of the genome of phi X174. It's a small virus that kills bacteria. We designed the pieces, went through our error correction and had a DNA molecule of about 5,000 letters. The exciting phase came when we took this piece of inert chemical and put it in the bacteria, and the bacteria started to read this genetic code, made the viral particles. The viral particles then were released from the cells and came back and killed the E. coli... And so, we think this is a situation where the software can actually build its own hardware in a biological system... And I've argued that we're about to perhaps create a new version of the Cambrian explosion, where there's massive new speciation based on this digital design.[57]

The implications of synthetic biology should captivate us, not because it is a matter of when, not if. When does this Cambrian explosion occur? The synthetic biology with which Venter has begun experimenting concerns new bacterial forms. But unquestionably, the leap to the human cell will only require the best and brightest to focus their attention away from bugs and onto the planet's principal specie.

The goals engineers will seek in the future are no different from the goals of engineers of ancient times: design and fabricate materials and structures into a logical order by which civilization fits together. In the future, the society will include a different kind of humanity based on a new form, which will be as novel as the wheel or the bridge was to the ancients or the telescope to Galileo or the microprocessor to Robert Noyce, Gordon Moore, and Andy Grove.

It was already one in the morning; the rain pattered dismally against the panes, and my candle was nearly burnt out, when, by the glimmer of the half-extinguished light, I saw the dull yellow eye of the creature open; it breathed hard, and a convulsive motion agitated its limbs.—Mary Wollstonecraft Shelley, *Frankenstein*, (1818)

WHERE WE ARE HEADED

In today's world, new research tools, bioengineered species, prosthetic devices, and state of the art computers overflow like cornucopias and transform society's health, environment, productivity, and war-making powers in ever-shorter periods of time. Some things are in their ascendency while others fade as they reach the nadir of their natural product cycles. Big "pharma" has perfected pills, such as Viagra® or Cialis®, that allow men to ready their body for sex, while women control their menstruation and prevent conception through the "pill." In the future, it is not inconceivable that these kinds of pharmaceuticals will be either replaced or combined with prosthetics delivered in nanobot vehicles, providing a permanent palliation rendering the pills unnecessary.

Future technologies, utilizing communication processes and other electronic apparatuses, such as semiconductors (smart RFID chips, micro lasers) can be extrapolated into future biomedical processes and apparatuses, with features that by today's estimates are within the realm of possibility, (i.e., not so futuristic as to reduce the exercise to one burdened by a heavy dose of speculation). Nano-biology, such synthetic DNA, is less developed than semiconductor technology and how quickly it matures and how it will be applied requires some assumptions and conjecture. Yet, this has always been true and should not deter a case to be made as to what the future may have in store. Certainly Alvin Toffler and prognosticators of his ilk occasionally get it right.

Moore's Law stands for the empirical proposition that over the history of computing hardware, the number of transistors on integrated circuits doubles approximately every 24 months and computer power doubles about every 20 months.[58] At this rate, it is anticipated that silicon based microprocessors will reach their capacity sometime towards 2030. However, it is projected that other computational technologies, such as molecular and carbon-based nano technologies will allow for scaling between the 10^{-7} down to the 10^{-9}—the large molecular level 20 to 100 atoms wide.

Joel Garreau, author of *Radical Evolution* argues that we are at an inflection point in history, where humanity as we know it will be decidedly different on the other side of the inflection. [59] He claims that we are engineering the next stage of human evolution through advances in genetics, robotics, information and nanotechnologies. These initiatives will alter our minds and metabolisms and in the course of the next fifteen years will become part of our everyday lives—making us smarter and defeating illnesses. But he warns that unrestrained technology may bring about the "ultimate destruction of our entire species."

The transformation will not be cataclysmic, but will happen over a long, protracted period, during which medicine will subtly change our normal routine. Most of the change will happen without our being aware of the artifacts and forms of technology that will slowly invade our bodies. In the early 1960s a radio pill known as the Konigsberg telemetry pill was ingested by patients for the purpose of recording internal temperature and then transmitting the data to a collecting antenna for use by medical professionals. [60] Jerome Schentag invented a pill in 1991 that when swallowed allowed it to be tracked electronically as it progressed through the alimentary tract and, upon reaching a specific site, was remotely triggered to release a dosage of medication. [61] Since then, small devices have attached video cameras to such pills to record the stomach and intestines, while other devices detect conditions and send signals to an electronic patch on a patient's skin that in turn transmits information to a computer. Much of this progress during the past 50 years has gone relatively undetected by the average citizen. What constitutes the technology, the small video processors, sensors, transmitters, the way we arrange the information on the computer chip, the information content on the chip itself, the manner of communication, or the very language of that communication? It is all technology, slowly, silently, finding its way into our anatomical cosmos.

As U.S. life expectancy goes well beyond the present 78.2 years—by several decades— traditional medicine, even the ever-improving pharmacology, will be replaced by implanted bio sensors, organ stimulators and pumps, utilizing the materials, electronics, computer processors and inter/intra-body communication networks to enhance or replace failed human organs. Physicians are already providing deep brain stimulation to adjust thinking and emotions. [62] In the future, it is conjectured that this therapy will be made available not as pharmacological products (such as Prozac) but as cybernetic packets that reprogram operating systems (the basal ganglia, thalamus, subthalamic nucleus, globus pallidus, internal capsule, nucleus accumbens) specific to brain functionality. The 2010 global market for microelectronic medical implants, accessories and supplies was an estimated US$15.4 billion and by in 2016 it is forecasted at US$24.8 billion. Though we can only survey this vast technological territory, seeing where we are currently situated is crucial to seeing what lies ahead. Material science (nano-technology, synthetic DNA, and non-genomic polymers), electromagnetic imaging technology to reduce the size of manufactured structures, and quantum physics technology will combine to allow continuing declines in component size and improvements in sensor sensitivity and computational power. These physio-chemical advances will complement the growth in the bio-chemical technologies and the unlocking of the inner workings of the brain, genome, and epigenome.

Implanted devices must be inert to biological rejection and as such are often encapsulated in a titanium shell. But titanium has drawbacks and in the future, ceramics, silicone on parylene coating, and glass encapsulation will prove more suitable especially as the need expands for reducing size, lowering power requirements, and improving radio frequency transparency to the outside world.[63] Some of the challenges for engineers over the next decade will be to ensure reliability by reducing errant leakages, attenuating external interference from other transmission devices, isolating them from the devastating effects of MRI scanners, and reducing electrical power consumption. Each of these areas is central to the efficacy of these devices and crucial in life sustaining situations.

We might classify these products broadly as those that deliver drugs, that sense physical conditions inside and outside the body (such as cochlear implants that sense sounds and convert them to nerve impulses), and electrically stimulate some aspect of our bodily process (brain implants, heart pacers). The FDA recognizes combination products as a special class; these associate electromechanical devices, such as pumps, with the delivery of a drug or biological product. At present, FDA combination products are either therapeutic or diagnostic, and there are no approved combination products for enhancing one's otherwise normal physiology.[64]

An implanted drug delivery system might include a pump, which stores and infuses a drug at a chosen rate through a pathway—typically a catheter—to the desired location.[65] Present incarnations of these products are relatively large: a typical implant for pacemakers or drug delivery can range from one-inch thick to one to three inches in diameter, respectively. Engineers anticipate that short term advances in pacemakers will scale this by fifty percent (the size of a large antibiotic capsule) and in the long ten to twenty year term, (as devices are custom fitted for a particular therapy) some devices will be downsized on the order of a grain of rice.

Today, implantable cardioverter/defibrillators, drug delivery systems, neurological stimulators, bone growth stimulators, and many other in-the-body devices significantly facilitate the treatment of a variety of diseases. But in an environment where embedded technology is ever more prevalent, treatment takes on a larger role than simply assuring that the body is healthy. With technology comes a service component. Someone needs to insure that batteries do not need replacing, that electrical probes are not corroding, and if dependent on computer programs, the code does not need updating to run reliably.

We are always looking to the future; The present does not satisfy us. —Our ideal, whatever it may be, lies further on.—Ezra Hall Gillett

EMBEDDED TECHNOLOGIES

The year is 2042, and Eve's grandfather, Dr. Robert Morgan, just had a birthday. He now prepares for his annual exam in the privacy of his home. Until the three-quarter turn in the twentieth century it was not uncommon for doctors to visit elderly patients where they lived. This changed, when in the latter quarter of the last century, elderly patients were seen by doctors almost exclusively in their offices. However, later this century, reactive medicine takes medical examinations to a new level where patients will simply login to their home computer and have all the benefits (and more) of a visit to the doctor's office. One of the added advantages to having a router and modem feature as one of the technologies embedded in the anatomy will be to allow access to medical care at any time, day or night, simply through a wireless Internet connection. By answering a series of questions to narrow down their symptoms, patients will first participate in their own diagnosis. An expert computer will determine the doctor best equipped to deal with the intake diagnosis. A worldwide grid listens in on millions of individuals like Dr. Morgan, feeding public health, searchable databases, used by the medical community including expert systems. Additionally Dr. Morgan's particular medical data will be updated and used by an expert system to make comparisons to his medical history and search medical references in the event that something is abnormal and requires a proactive response.

Dr. Morgan, being an early adopter and also in need, sports many prosthetics, without which he may not have survived to the ripe age of 100. The reactive server Dr. Morgan accesses takes an initial inventory of the installed technology:

1. A capsule-sized pacemaker embedded in the interior wall of his heart to keep his occasional arrhythmia in check (heartbeats that are too fast or too slow);

2. A processor that manufactures and supplies insulin to keep his sugar levels in check;

3. Two cochlear implants buried near each ear drum so he may hear his favorite operas again;

4. Two artificial hip joints and an artificial knee with microprocessors that adjust the formability of the prosthetic as it wears out;

5. A deep brain stimulator to alleviate trembling hands as he battles with signs of Parkinson's disease;

6. A second deep brain stimulator to ease a long term depression that followed after losing his wife in an accident;

7. A retinal implant after losing his sight to macular degeneration;

8. After the retinal implant, his family doctor suggested that he have installed a router/modem—a small computer for specialized communication that can hook up more effectively with the Internet and rout signals to each of the nine embedded microprocessors.

As indicated Dr. Morgan has always been an early adopter, and as he knows, the technologies he has now installed have been around, (albeit in the early years as experimental), for the better part of 30 years—or since 2012! But let us venture into the speculative future and see what may be ahead for Eve and her children.

Before Dr. Morgan lives out his expected lifetime into the early 2070s, he, his granddaughter Eve, and her children will see breakthroughs in nanotechnology. These will take place in two stages. In the first stage—expected in the next few years—engineered materials, such as glass beads with gold coatings, will be used to passively penetrate the cell—particularly cancers cells—and due to their absorption of infrared wavelengths, zap the disease when the beads are irradiated. Many products along these lines will follow. In the second stage—expected in the next few decades—these materials will function dynamically as roving nanobots and molecular computers that will blur the boundaries between natural and synthetic molecular systems. Some of these systems will operate autonomously to keep biological processes highly tuned. Others will communicate inner body data to larger external data bases where merging of data from all other similarly bio-equipped individuals will insure everyone is linked and synced-in with the latest trends in wellness remediation. Advances in bioscientific research will extend life through genomics, epigenomics, and proteomics, where delivery systems of older drug prescriptions will be replaced by digital applications or APPs transmitted over bio-information highways to all individuals possessing certain computer-driven, bio-operating systems that can interface directly with the anatomy's subsystem (more about this later).

Outside the body, the social sciences will cultivate social "memes": ideas, behaviors, and modes of living that spread from person to person within a culture for the purpose of forming a collective IQ. Imagine that it amplifies what anyone person knows by affording instant access to information. These social memes will change the way we transmit information and will revolutionize what we know, how we are educated, choose friends, choose lifelong partners, and contribute to the architecture of what Teilhard described as a "gigantic psychobiological operation, a sort of mega-synthesis, the 'super-arrangement' to which every individual will at once be a collective subject."[66]

Some of Dr. Morgan's prosthetics use computer processors, while others combine polymer technology, such as polymeric smart materials, that have the ability to return from a deformed temporary shape to their original shape in applications such as artificial spinal discs. In the non-electronic realm are combination products such as coated or impregnated drug or biologic articles, for example, drug-eluting stents; pacing leads with steroid-coated tip; catheters with antimicrobial coating; orthopedic implants with growth factors. None of these were mentioned in the inventory that Dr. Morgan carries for the balance of his life, but they will certainly be in every physician's medical bag in the near future. The technologies having the greatest sociological impact link in-the-body technologies (intra-computer-communication systems) to outside-the-body data processors. Here, I refer to a combination product as that which resides in the body in the form of a computer control system for: drug infusion, electrical excitation, or electronic sensing as coupled to one or more of an electrical circuit, mechanical device, or drug delivery apparatus.[67]

Memory, wit, fancy, acuteness, cannot grow young again in old age;
but the heart can.— Jean Paul Richter

BODY MEET COMPUTER—COMPUTER MEET BODY

The heart as it beats 100,000 times per day (about
2,500,000,000 beats in the course of 70 years) is the most
vital human organ since any catastrophic breakdown
usually spells sudden death. It is also the part of the
anatomical matrix where electricity plays a significant role
in how the heart maintains its timepiece-like
synchronization.[68] Its electrical signals spread wave-like
from the top of the heart muscle, over the surface of the
heart, to the bottom causing the muscle to contract and
pump blood. Each signal begins in a group of cells called
the sino-atrial node. The heart is divided roughly down the
middle by a separation wall called the septum. Each half
has an upper chamber called the atrium and a lower
chamber called the ventricle. When the atria contract, they
pump blood into the heart's ventricles or lower chambers,
and when the ventricles contract they pump blood to the
rest of the body. The combined contraction of the atria and
ventricles is the electrically stimulated heartbeat we feel
inside our chest cavity.

The most common problem with the heart is
arrhythmia, i.e., heartbeats that are too fast, too slow, or
that have an irregular beat. During an arrhythmia, the
heart may not pump enough blood causing fatigue,
shortness of breath, fainting and in the worst case death.
The pacemaker, a small computer with complicated
software, can for the dysfunctional heart, insure that the
ventricles contract normally and that the atria are not
quivering (referred to as atrial fibrillation) by synchronizing
electrical signaling between the upper and lower chambers
of the heart; and in some cases supplying the lower
chamber with an electrical signal if there is a disconnect in
the electrical circuit between the upper and lower
chambers.

In 1886 Walter Gaskell discovered that the heart's sinus venosus was the area of first electrical excitation.[69] Forty-four years later an electrical circuit was designed and located outside the body to supply a regular current impulse through the chest cavity wall to stimulate a dysfunctional heart.[70] In 1952, Paul Zoll, a Boston cardiologist, developed what many generally regard as the first viable pacemaker when his device detected a missing heartbeat and sent electrical shocks through the chest wall and muscle below. This form of "electrocution" was painful, burning the patient where the electrical current was injected. In the mid 50's the newly invented transistor meant that the pacemaker could be made small enough to be implanted within the body and, by 1960, Clarence Greatbatch received a patent for a pacemaker that used transistors. Other problems with the first pacemakers had to do with short-lived batteries, but this problem was solved a decade later when the longer-lasting lithium batteries became standard. As will be discussed below, pacemakers continue to shrink in physical size and power requirements to the point where energy scavenged from the body's metabolic processes eventually will keep the next generation pacemaker's micro-circuitry powered almost indefinitely.

A modern pacemaker consists of a computerized generator, sensors, and one to three wires that are each placed in different chambers of the heart. The sensors detect the heart's electrical activity and send data to the resident computer. If the heart rhythm is abnormal, the computer controlled generator sends electrical pulses through the wires to the heart. Newer pacemakers can monitor blood temperature, breathing, and adjust heart rates according to physical activity. The computer memory also records the heart's electrical activity in the form of histograms, which a doctor may read during an examination to determine if the heart had undergone, since the prior examination, any irregularities (e.g., missing beats, endogenous atrial currents that correspond to the P wave in an electrocardiogram). Through a two-way radio communication channel connecting to a computer in the doctor's office, an electrocardiologist can also change the operating parameters of the pacemaker's computer.[71]

Aside from the data that the pacemaker collects, there remain opportunities to diagnose certain conditions from auditory vibrations (i.e., heart sounds); these that may reveal anomalies in pumping, such as a heart murmur or mitral regurgitation. Two normal heart sounds, commonly known as the S1 and S2 sounds, often can suggest the onset of ventricular systole or the onset of ventricular diastole, respectively. In the near future the pacemaker will be connected through something like Bluetooth technology from a local computer to a central server that can analyze any abnormalities and report to the physician. The patient will be alerted to seek medical attention, unless it more along the lines of an emergency, in which event, he or she will be alerted to wait for a first responders who will have been notified by the system.

Everything in the body is traceable to an electro-chemical reaction, whether it is the sound we hear, the light we see, moving our thumb, and all that which we do not sense, the release of digestive enzymes, the movement of food through our gut, the thoughts we have as we read this text. So it is not surprising that electrical stimulation, such as provided by pacing devices, similar to the pacemaker, are used to reduce the effects of Parkinson's disease, to correct obsessive compulsive disorders, to stop essential tremors, reduce depression, dystonia, epilepsy, gastroparesis, obesity, bowel disorders, interstitial cystitis, urinary incontinence, and chronic back pain.

Medtronic, a leader in pacemaker technology, in 2010, announced a totally self-contained intracardiac pacemaker device.[72] Rather than the computer and battery being installed in a package the size of a silver dollar, and implanted beneath the skin under the clavicle near the shoulder, the new device will be lodged completely within the heart. Instead of leads with small harpoons that now run from the pacemaker to the interior of the heart, the new device has small grappling hooks that fasten themselves into the interior heart wall. Implantation of the new device requires snaking the pacemaker—a container the size of an antibiotic capsule—through the femoral vein, the one doctors use for angiograms, negating any surgery and thus carrying a low risk of infection. The device has a miniature radio for communicating with a nearby smartphone for either the patient or the physician to interpret the signal traces. But, the story does not stop here. A spin off of the capsule soon will be installed to monitor pulmonary pressures in the lungs, or beneath the skin above the area of the heart to monitor heart rhythms. The signals will also be transmitted to a smart phone. Within the next five years, the devices mentioned here will be reduced from the size of an antibiotic capsule to the size of an aspirin.

As indicated, pacemakers now installed in millions of patients, incorporate radio technologies to communicate to the outer world to receive updates and alert physicians of serious conditions. As pacemaker-type technology, and in-the body-computer devices generally, decrease in both size and power requirements and interface with smartphones, telemedicine issues will arise over who can access to the in-the-body computer and for what purpose. Medical ethicists and policy makers will have to address the extent to which the central computers that communicate with wide swaths of the patient community, through in-the body-computer devices, will require regulation and institutional controls to assurance reliability, privacy and security, and ultimately patient safety.

Technology presumes there's just one right way to do things and there never is.—Robert M. Pirsig

REENGINEERING PERCEPTION

Energy-packing information in the forms of sound, sight, smell, and touch pours through more than 37 sensory inputs on their way to the brain. Some of these sensory inputs employ the over 260 million visual cells, the almost 50,000 auditory cells, or the approximately 78,000 receptor cells for touching, smelling, and tasting. Cells too tiny to see with the unaided eye scan the environment like omni-antennae reaching for the truth, separating and integrating the raw uncut data that flows through its central nervous system. The stream is incessant. Some is incomprehensible chatter, while other pieces compel us to focus because they affect our needs, beliefs, and survival. But what if in addition to the noise and the essential data, we were supplied by yet another source that discreetly feed us data directly to the cerebral cortex? It may sound like science fiction, but we may be closer than imagined.

Beasts of all stripes compete for the limited bounty the world produces; so for any species the ability to hear, see, and smell accurately and quickly is undoubtedly advantageous to survival. Anthropologists believe that intelligence was made possible through our ability to sense the environment, to physically grasp objects within reach, and to make intelligible utterances. These abilities manifested in a spoken language that richly expressed abstract schemes. This last feature led to the creation of physics and mathematics, which brought about computers, lasers, satellites, wired and wireless transmission and programming. We stand on the cusp of a revolution in in-the-body communications, to and from the outer world, where what happened with the advent and proliferation of the Internet will be repeated in the Intranet of our bodies.

In redesigning humans to improve the odds of survival or enhance a sense of well-being, technologists will engineer perceptions that offer light when it is dark and offer sound when there is silence, effectively bypassing the five senses. This will link the outer world directly to the brain and will set us upon an evolutionary course change.

For some time after one's birth, the cerebral cortex remains plastic allowing a vast network of neurons to shape itself into an analog of experience. At any point in time, an organism that learns through experience exists in an antecedent context: a constant feedback that conditions an adaptive, unique persona. What we believe depends on this stream of data giving rise to our fears and motivation for flight, biases, prejudices, and preferences, refined by repeated experience and inferences drawn therefrom.

What if through science and engineering, we might alter this experiential conditioning through electrical code based brain stimulation? What if our aim were not therapeutic, but to augment healthy perceptions through such artifices as virtual reality? This would help us learn a new subject or skill—possibly even as we attend to another—to learn while sleeping, not a chaotic dream, but imposed as an image on our visual cortex.

Pharmacology and food science already do this while we are awake using psychotropic drugs (e.g., Adderall and Ritalin) that improve mental functions such as memory, intelligence, motivation, and concentration.[73]

Electrical code-based brain stimulation could change our social reality—that is, what things mean and what things satisfy our desires—, leading to more opportunities for learning, better health, and greater security and personal and economic freedom.

Can we develop systems that anatomically integrate technology and the brain for that purpose? The answer is not a matter of if, but a matter of when. To achieve this, it would seem that we need to penetrate the mind directly—, something scientists are quickly leaning how to do with great success.

As the fire-fly only shines when on the wing, so it is with the human mind—when at rest, it darkens.—L.E. Landon

THE DRUG—ELECTRONIC TRADE-OFF

We are large electricity generating machines, powered as it were by charged atoms that determine what we see, hear, taste, feel or do. Bioelectricity is a result of ions, as they pour over organ surfaces or travel through cell membranes and along axons, the "wires" that constitute our nerve cells. Ions are charged particles similar to the electrons that flow in electrical wires, but that move slower than the electron currents that travel down wires at 186,000 miles per hour. And so it is no surprise that scientists and engineers look for electrical solutions to medical problems from diabetes and heart disease to pain management and changing dysfunctional brain patterns.

The neocortex contains over four million neuronal modules. Each module has a few thousand neurons and is vertically oriented across the cerebral cortex, which hammer out signals acting like something roughly equivalent of piano keys.[74] In response to brain stimulus, these "keys" send spikes of electrical currents having the parameters of intensity, duration, rhythm, and simultaneity, sounding out an uncountable repertoire of spatiotemporal tunes.

At the "component level" of our nervous system is a nerve cell—analogous to an active electronic circuit containing an input signal, a transforming processor, and an output signal. An input from another neuron or outside stimulus is received by a dendrite, which has many branches resembling roots of a tree. The dendrite serves to combine the inputs and transmits a representative electrical impulse through the cell body and along a communication channel called the axon. The axon is analogous to a wire that then outputs an electrical signal to as many as six thousand synapses. Researchers have long understood the electrochemical environment in which neurons pass on sensory data from dendrites to synapses. It is sometimes electrical, but oftentimes is more complicated. A gap exists called the synaptic cleft. When a spike arrives, it causes a chemical release through tiny vesicles. These chemicals diffuse into the gap between the synapse and the next receiver—perhaps another cell. At the atomic level, the process is a consequence of gates opening and closing, thereby allowing ions to flow in and out of membranes on the postsynaptic side of the synapse. The effect is to create a voltage potential and thereby create current flow. In short, the impulse of current down the axon causes a chemical to be released, which in turn causes a physical effect such as a muscle contraction.

Needless to say, the electrical nature of the central nervous system has led to the idea that electrical stimulation can be used to alter the mind's thought pattern. In a destructive way, this is what convulsive shock therapy does; however, in a less destructive way, pacing signals can be used to pace the basal ganglia, the primitive deep structures at the base of the brain that route thought and senses either to the cerebellum or the spinal cord. Pacing overrides errant brain signals and has been effective for Parkinson's and dystonia, obsessive compulsive disorders and drug addiction. Pacing itself is rather crude, since it does not cure the underlying malady, but scientists are currently applying electronic signal theory to alter brain patterns with greater degrees of finesse and efficacy, that will benefit various conditions that respond to the electrical codes that simulate the brain's "software".

Much of the research to date has been done in the area microstimulation to identify the functional significance of groups of neurons. Research in 2006 applied microstimulation to rats' prelimbic and infralimbic subregions of the medial prefrontal cortex to test fear-induction, concluded that the prelimbic subregion excites fear while the infralimbic subregion inhibits fear.[75]

Microstimulation is also being used to deliver signals to circumvent damaged sensory receptors or pathways. For example, as discussed below, the stimulation of the thalamic or visual cortex has successfully created visual images or phosphenes restoring partial vision in the blind. Other neuron stimulation will be applied to bladder prostheses, cochlear prostheses, and brain-stem auditory prostheses. In the future, the application of electronics may replace drug therapies, such as those used to improve cognition. Recent experiments with the effects of the drug Ritalin serve as the opening.

Ritalin has been used to treat attention deficit hyperactivity disorder, but in low doses it appears to effectively boost cognition. In 2008, University of Wisconsin-Madison psychology researchers, David Devilbiss and Craig Berridge, were working under grants from the National Institute on Drug Abuse, the National Institute of Mental Health, and the UW-Madison Discovery Seed Grant Program. They reported that Ritalin fine-tunes the functioning of neurons in the brain's prefrontal cortex where attention, decision-making, and impulse control are managed. [76]

Using an electronic system for simultaneously monitoring many neurons, the scientists observed both the random, spontaneous firings of prefrontal cortex neurons and their response to stimulation of the hippocampus, a central pathway to the prefrontal cortex. Acting like miniscule sensors, electrodes recorded a spike every time a neuron fired. According to Devilbiss, analyzing the complex signal patterns was, "Similar to listening to a choir, you can understand the music by listening to individual voices... or you can listen to the interplay between the voices of the ensemble and how the different voices combine."

The scientists found that cognition-enhancing doses of Ritalin had little effect on spontaneous activity, but the neurons' sensitivity to signals coming from the hippocampus increased intensely. Berridge said that, "[y]ou're improving the ability of these neurons to respond to behaviorally relevant signals and that translates into better cognition, attention, and working memory."

During their study, the researchers were able to tune or time-synchronize the entire chorus of neurons. When groups of neurons resonated as one, Ritalin reinforced this coordinated activity, while lessening any uncoordinated activity.

Electronic technology is well suited to intervene in therapies that conventionally use drugs, largely because the neuronal process is one based in electronics, and because the field has a two hundred year history of developing technology that can focus electrical signals in the form of impulses and waves with a high degree of precision and control. We are at the inception of a significant turn in the trade-off between drug therapy and electronics.

... Seeing is not as simple as you might have thought. It is a constructive process in which the brain responds in parallel to many different 'features' of the visual scene in attempts to combine them into meaningful wholes, using its past experience as a guide. Seeing involves active processes in your brain that lead to an explicit multilevel, symbolic interpretation of the visual scene.— Francis Crick, *The Astonishing Hypothesis* (1994)

ELECTRONIC ART OF SEEING

Our senses situate us within the world; it is left to the brain to establish what kind of world it is. A significant part of how we perceive that world comes from the eyes' ability to form images over a narrow band of wavelengths of light that emanate from objects within our field of vision and, in cooperation with the brain, direct our attention. In the human eyeball, there are three layers referred to as the sclera, the iris, and the retina. The sclera forms the white of the eye—the interior part of which becomes clear and transparent to form the cornea. The next layer is the iris, the colored portion of the eye, which is comprised of muscles that control the amount of light that passes into the eye from the outer world. The hole created by these muscles within the iris is the pupil, behind which a lens is held in position by various ligaments in the space between the iris and the retina. In a space referred to as the optic-cup, the retina and its more than 100 million photoreceptors begins the process of turning light into information. To focus, the incoming light falls on the retina, but on the way, the rays must be refracted or bent through the lens with the help of the cornea (its aqueous humor and vitreous body all taking part in the refraction). The eyes must also adjust to the intensity of light, so eye muscles cause the pupil to dilate or constrict, adjusting the amount of light that reaches the retina.

In addition to converting an image field into neuro-impulses, the normal eye must also move in various planes. The socket of the eyeball (referred to as the orbit) is about one inch in diameter and contains the eyeball, which projects forward beyond its opening. The eyeball performs six muscle movements through the tension and relaxation of seven major muscles. The superior rectus muscle and the inferior rectus muscle are both inserted on the upper and lower surfaces of the eyeball. The superior oblique muscle runs through a facial sling or pulley and is then attached obliquely to the upper surface of the eyeball. The inferior oblique muscle arises from the bottom of the front of the orbit and runs obliquely on the inferior surface of the eyeball. The medial rectus muscle and the lateral rectus muscle are attached to the surface of the eyeball. The movements of the eyeball are due to the relaxations and contractions of the muscles, which provide for turning the pupil up, down, sideways and even in rotation.

The eye serves as a major form of direct input to the brain and, as technology moves in the direction of enhancing sensory experience, this feature would seem to have a strategic advantage toward fulfilling that objective. Enhanced visual experiences are of interest to military tacticians, certain industries, doctors who want to see more than is possible with the naked eye, and permits gamers and others in search of new kinds of amusement. Technology that interface directly with the visual framework also has significant medical benefits for individuals who are blind or at risk of losing their sight due to retinal degenerative diseases, congenital defects, or accidents. It is through research to ameliorate blindness, that other applications, such as interfacing information systems through the optic nerve or visual cortex will be realized.

Since the late 1960s biomedical scientists have striven to restore visual perceptions to irreversibly blind patients using implanted electrodes for electrical stimulation of the eye, optic nerve, and brain. [77,78] Of these, electrical stimulation of the visual cortex is the only approach has so far to allow a blind person mobility. This was achieved by connecting a television camera directly to the visual cortex and restoring some level of visual perception.[79] Although prosthetics have been largely limited to demonstrating the perception of spots of light and high-contrast edges, researchers recently used the retina's neural code for driving retinal stimulators and found it had the potential to restore normal vision.[80] Advances are approaching at lightning speed.

At approximately 250 microns thick, the retina has layers consisting of thin sheets of photoreceptor cells called rods and cones. These respond to incoming light by firing over a million ganglion cells and converting these to a sequence of electrical spikes, which differ according to the intensity of light falling on the retinal surface. The retina is attached to the optic nerve, consisting of more than one million axons that transmit several hundred electrical spikes per second to the visual cortex via the lateral geniculate nucleus—, a small part of the thalamus. These spikes vary in frequency similar to pulse coded modulation used in communication systems—, and they carry the data or information necessary to perceive the outside world.

When blindness occurs, it is often because the retinal photoreceptors have been compromised, due to diseases such as retinitas pigmentosa and macular degeneration. Nevertheless, despite degeneration of the retinal architecture, the inner retinal neurons are often amenable to electrical stimulation. By activating these surviving cells, signals are relayed to the brain; these cause a visual sensation called "phosphenes." The mechanical, electrical and chemical parts of the eye and the eye's support system are fairly well understood, and most of the science to restore vision is centered on developing electromechanical devices,—such as electrode arrays—for retina neural stimulation. Most current prosthetic devices that might stimulate the visual cortex through a dysfunctional eye depend on an intact optic nerve. However, in instances where the optic nerve may be damaged, similar electrode technology is being used to excite the visual cortex, directly. Once the image is impressed on the visual cortex, patients have come to visually perceive variations in light, shape, color and movement in the imaged field.

The idea of electrically stimulating retinal cells to produce flashes of light or phosphenes has been known for quite some time.[81] The modern era in retina prosthetics began when Brindley and Lewin (1968) used eighty cortical surface electrodes in a patient to study electrical stimulation effects through the visual cortex. They found: (1) consistent shapes and position of phosphenes; (2) that increased stimulation pulse duration made phosphenes brighter; and (3) that there was no detectable interaction between neighboring electrodes which were as close as 2.4 mm apart.

Some forms of blindness involve selective loss of the light sensitive transducers of the retina. Other retinal neurons remain viable, however, and may be activated in the manner described above by placement of a prosthetic electrode device on the inner (toward the vitreous) retina top layer (epiretinal layer). This placement must be mechanically stable, minimize the distance between the device electrodes and the visual neurons, control the electronic field distribution and avoid undue compression of the visual neurons.

Implantable arrays of planar electrodes were first used to record signals from cultured cells in 1972 by C.A. Thomas, Jr. and his colleagues.[82] The experiments used a 2 x 15 array of gold electrodes plated with platinum black, each spaced 100 µm apart from each other. To bionically create images, two types of implantation are currently being tested: planar types and 3D "bed of nails" array. Using photodiode technology to capture the light, the electrical signal is coupled to either the retina's top, epiretinal layer, or beneath the retina's subretinal layer. In a device that interfaces to the top layer, the array is 4.2 mm square with 100 silicon micro-electrodes placed 0.4 mm apart. One subretinal device is 2 mm in diameter with 5,000 microelectrode-tipped silicon-based photodiodes consisting of 16 platinum discs arranged in a 4 x 4 square array and powered by ambient light. A single 25 micrometer diameter platinum wire is attached to each disk and the assembly is encapsulated in silicone rubber. A third approach interfaces directly into the cortex. This device utilizes a 3D needle array technology (referred to as the Utah array) that "plugs" into the brain.[83,84]

In one of the most advanced products to date, the Intelligent Retinal Implant System [TM] prosthesis bridges and replaces the defective information processing function of the actual retina in patients with retinal degeneration.[85]

It may enable the blind to recover visual perception and a sense of orientation in unfamiliar surroundings.

Another product, the Learning Retina Implant System™ artificial eye consists of three components: a retinal stimulator (implanted in the eye), a visual interface and a pocket processor.[86] The visual interface resembles a standard pair of glasses but incorporates several electronic components: a camera to capture images from the wearer's surroundings and other components for data communication with the pocket processor and the retinal stimulator. Image information is processed by the pocket processor and translated into stimulation commands to the retinal stimulator inside the eye via wireless transmission. They are then transformed into electrical pulses by a microchip in the retinal stimulator to electrically signal the retina and provide selective visual perception.

Tracking the latest progress in the creation of electronic solutions to improving sight, we find that scientists are moving rapidly with the development of the biological eye using stem cells. Organogenesis, the formation of our organs, is a self-organizing process carried out in three-dimensions where changes in the external and internal shape of bundles of cells fold inward and outward until the desired structure emerges. In a recent study scientists witnessed the optic cup, upon which the retina forms, spontaneously materialize from human embryonic stem cells.[87] In another study, scientists grew "to approximately two millimeters in diameter, with a single-layer retina epithelium becoming, as in the embryo, a stratified structure containing all six categories of cells found in the postnatal retina." [88] These developments are the initial steps in the creation of bioengineered retinas, expected, as will their electronic counterparts, to eventually improve the lives of the millions with such sight diseases as macular degeneration.

We have seen that every sound, and every succession of sounds, can be represented by a curve, and our first problem must obviously be to find the relation between such a curve and the sound or sequence of sounds it represents—Sir James Jeans

COCHLEAR IMPLANT SYSTEM

Sensorineural or nerve deafness is caused by missing or damaged hair cells in the cochlea, which transduce acoustic signals into nerve impulses. Mild to severe sensorineural hearing loss can often be helped with hearing aids or a middle ear implant, but something more sophisticated (like the cochlear implant) is the only solution for profound hearing loss. Approximately a quarter-million people worldwide have received cochlear implants, a device that is surgically implanted, which bypass the ear's hair cells to deliver electrical stimulation to the auditory nerve fibers and allow the brain to perceive sounds such as speech and music, even in noisy surroundings.

Cochlear implant systems have typically consisted of two parts: one is worn behind the ear and consisting of a microphone/processor; the second part, a stimulator/receiver is implanted within a recess of the patient's temporal bone. The external component converts detected sounds into coded signals and transmits them transcutaneously to the implanted stimulator/receiver unit. It is anticipated that in the future the entire system will be implanted within the head of the user. The implanted unit typically includes an antenna receiver coil that receives the coded signal and outputs a stimulation signal to an intracochlea electrode assembly which applies the electrical stimulation to the auditory nerve producing a hearing sensation corresponding to the original sound.

For the power of Man to make himself what he pleases means, as we have seen, the power of some men to make other men what *they* please.—C.S. Lewis, *The Abolition of Man* (1943)

ACCESSING THE DEEP—SENSORS AND CONTROLLERS

Cells express physical properties that engineers recognize as electric charge, viscosity, modulus of elasticity and volumetric density. These are all measurable properties, which when coupled through sensors can be tied into a computer for analysis. The convergence of computing and biology—that is a microprocessor embedded into our physiology at various sites—can form an element of a computing network that measures and analyzes the logical states of our anatomical subsystems. In this way the molecular properties of cells can be computationally analyzed. Soon, planted deep into our anatomy a living laboratory will be managed by hybrid computers: partly digital, partly analog; partly silicon and/or carbon, partly molecular/DNA; partly electronic, partly photonic, partly fluid, and partly quantum mechanical. Through the power of algorithms of the type used in stock market analysis, embedded bio-computer systems connected to sensors will be capable of predicting disease and regulating the biological function of the organ to which it is assigned.

Most biomedical products incorporate some type of sensor for extracting the physical parameters needed to make measurements that in turn result in some kind of action. Some sensors are passive—such as a thermocouple that may be used in an electronic thermometer. It requires no energy source, since the thermocouple converts heat into voltage. Other sensors, such as piezoelectric ceramic sensors, convert mechanical pressure into voltage. For example, by using sensors fabricated from piezoelectric ceramic material doctors can measure heart contractility, but most of these sensors also require an external power source like batteries, which need regular replacement. In one recent innovation, an implantable piezoelectric sensor used an attachment to hold a sensor element in contact with tissue such that when the element bends in response to the flow of fluids and adjacent tissue the bending produces a usable voltage.[89] In some embodiments, the attachment device holds the sensor in direct contact with blood flowing through a ventricle of the heart. In another embodiment, the sensor is in contact with an epicardial or endocardial surface of the heart or a surface of a pericardium. A radio-frequency transmitter may be connected to electrodes to transmit an output signal regarding the rate and magnitude of the heart's contractility.

Other sensors, such as enzyme biosensors, can be inserted into the body to continually sense glucose levels in managing diabetes or detecting hypoglycemic events. For nearly 12 years subcutaneous needles have contained sensors to measure glucose. The sensor attaches to a needle that is removed after sensor insertion. The sensor generates tiny currents that can be read off by an external meter. Some of these products can measure glucose every 10 seconds and report glucose concentration at scheduled intervals. The external receiving unit is the size of a small cell phone and provides patients with auditory or vibratory alerts when glucose levels exceed programmed thresholds.

Recent innovations connect to the Internet via cuffs or gloves and transmit vital health conditions, such as EKGs. Other systems initiate voice and video links between a patient and medical professionals. Soon to reach the medicine cabinet is the MinION ™ DNA disposable USB-like sequencer that connects to a home computer and sequences an individual's DNA. Close by is the Zio ™ patch worn near the heart that reports irregular heartbeats through an iPhone app. For patients with or at risk for glaucoma, a contact lens with an embedded microcomputer will soon be available to measure eye pressure and will send the results directly to an ophthalmologist.[90]

In August 2012, the FDA approved an ingestible "Digital Pill" to monitor medication adherence in patients.[91] The pill contains a microchip to record and transmit medical data. Using a silicon wafer, about the size of a grain of sand, trace amounts of magnesium and copper generate electricity when the pill contacts digestive fluids. At that point the pill is conditioned to send its data to a skin patch, which in turn transmits the acquired data to a health care provider via a mobile phone. Although today's product records the time the pill is taken, patient's heart rate, body position, and temperature, it is slated for more complex sensing and transmission in the near future.

Biological sensors in the form of neurochips have been being implanted in animal subjects for many years. In the 1999-2000 timeframe, a brain-machine interface, consisting of hundreds of hair-thin microwires was implanted in a rat's brain and captured electrical signals when the animal moved. Most of the research was not on anyone's radar until a microchip was implanted into a chimp to control its neuro-motor; this allowed the chimp to control a computer curser by using its thoughts.[92] In 2005, a tetraplegic became the first person to control an artificial hand using a brain-computer-interface as part of the first nine-month human trial of Cyberkinetics Neurotechnology's 96-electrode BrainGate-chip-implant. The chip implant, which was installed in the patient's right precentral gyrus (area of the motor cortex for arm movement) and allowed the individual to control a robotic arm by thinking about moving his hand to control a computer cursor, the lights, and a TV.[93]

In 2007, Duke University Center for Neuroengineering recorded the activity of more than 200 cortical neurons in a rhesus monkey as it walked on a treadmill. It transmitted its corresponding electrical activities from Durham, North Carolina to a robotics' laboratory in Kyoto, Japan via the Internet. At the robotics' laboratory, a robot detected the electrical activities and began to walk as well. The implications are enormous, as it leads immediately to speculation that such interfaces could control distant surgical activities, control robots for dangerous assignments, or in the future, for an individual to control other devices implanted into the anatomy,—all through nothing more than thought.[94]

In May 2012, before network television cameras, Cathy Hutchinson a 58-year-old woman paralyzed by a stroke and unable to move her own arms or legs, sipped her cinnamon latte with the help of a mind-controlled robotic arm and an implanted sensor about the size of an 80 mg aspirin. The BrainGate device bypassed nerve circuits and replaced them with wires that ran outside Hutchinson's body.

For the scheduled 2014 World Cup in Brazil, Miguel Nicolelis and his team from Duke University are planning to outfit an exoskeleton on a paralyzed teenager who is incapable of using her legs. She will open the ceremonies by using her thoughts to raise herself from a sitting to a standing position; she will walk to a soccer ball, hook her prosthetic foot under the ball, and boot it into the space of a stadium. This remarkable feat will be witnessed by more than a billion viewers around the world.[95]

Within the next twenty-five to thirty years, nanotechnology is anticipated to reduce the size of sensors and stimulators to a size that will allow blood cell-sized apparatuses to interact directly with neurons and other cells and organs. Until then, scientists will make progress through using improved manufacturing techniques and the ever-expanding comprehension and use of quantum mechanical computers.

Once a patient has installed a medical device that regulates and controls some aspect of their anatomy to maintain them healthy, vibrant and frankly to assure life, they are less likely to consider the device alien or have an aversion to it.

It is the tension between creativity and skepticism that has produced the stunning and unexpected findings of science.—Carl Sagan

ARTIFICIAL BRAINS

During the course of our daily lives, trillions of creative impulses flow into our sensory pathways and are impressed upon millions of nerve endings. Each represents a separate experience. We respond immediately to some while others go apparently unnoticed only to reemerge years later in something said or felt—perhaps as an idea that recreates our world. Uncountable quanta, waves, and bits bore their way into the mind's cosmos to be later reclaimed; in the vastness of this inner space, in the folds and valleys of a cellular terra firma we call the cortex, most fall still, and lies buried forever, limiting what even the wisest and most intelligent among us can fathom. The greater amount of what does not get lost must navigate the depths of uncharted gray matter, to carve out new streams of predilections, prejudices, biases, and expectations. In full regalia she emerges an "idea"—a construct of consciousness, of data flowing into and out of a tangled biological bundle: the human brain qua mind. The mind is an organ of tumult and turbulence, a construct of social reality, where every so often erupts the idea that concretizes the soul or saves the world. Who, we might inquire, owns this genius? And is it possible to improve upon this through the virtuosity of science?

By 2020, computers will perform 10^{16} calculations per second and have storage capacities of 10^{15} bytes, which come close to the processing power of the human brain, but the right model and software will be required if the brain itself is simulated. The brain has functionality that controls not only conscious thought, but directs and controls sensory inputs and outputs as well as the metabolic systems that regulate the entire anatomy. If we consider merely the ability to recall recordable information in some structural way, then we might look to how the latest search engine technology retrieves information. IBM Watson-like artificial intelligence allows computers to do what any efficient mind can do. Alan Turing predicted that a computer would eventually pass for another human when an interviewer would not be able to distinguish the computer's response from that of a human. [96] It may well be that the computer has passed that test.

Henry Markram runs the Blue Brain project that uses an IBM supercomputer optimized for large-scale simulations of the brain's circuitry. By 2014, Markram claims that his team will have simulated a rat's brain totaling 100 million neurons and one trillion synapses, and astonishingly by 2023 will have simulated a full human brain.[97] Others estimate that it will take considerably longer, and no sooner than 2050, when the anatomical models and software will effectively have reverse-engineered the human brain. The anticipation is that it will be adept in making intelligent decisions based, not on reductionist learning theories, but rules of thumb, educated guesses, intuitive judgment, or common sense.[98] Although predictions vary as to when this will precisely occur, a considerable base of support exists for the notion that the human brain will be simulated during the first or second quarter of this century.

Let us return to the question of whether science will someday have technologies that employ a direct virtual reality connection to such a computer that will allow us to store and retrieve information or receive input for learning and experiencing—without using any of our five senses. It has been found through artificial retina research and brain studies, that synapses and dendrites can be artificially activated by passing pulses of electrical current down axons, causing changes in electrical potentials across the visual neuronal membranes. Based on this mechanism, it is possible to input coded sensory data in the form of sequences through a prosthetic connection to the axon. Certainly this idea has been successfully demonstrated in retinal and cochlear implant systems.

For obvious reasons much brain research has to do with using one's thoughts to control events outside the body in a form of psycho-kinesthesis, where those who are paralyzed can gain a measure of control over their situation simply by thinking through what they desire. But what about the outside environment, where computers might control the individual by shaping what the individual "knows" or "sees?" This is essentially what engineered retinal and cochlear systems try to accomplish. In the future, bioengineering processes, which utilize computers, will influence thoughts and memories without recourse to our five senses, by interfacing directly with the brain. This interfacing will revolutionize how we experience realms other than the natural world—realms more akin to virtual reality.

Since the late '50s, researchers have known that neurons located in the cortex respond to stimulus received from the thalamus as it relays signals from specific parts of the visual field.[99] When a neuron's receptive field is sufficiently excited, it sends a series of electrical spikes down its axons (or body), in response to color, movement, and size of objects in the visual field. When there is no activity in the field, the resting or quiescent frequency of the spikes ranges between 1 and 5 Hertz. When excited by an event, the rate increases to typically 50—100 Hertz, and for short periods, may reach 500 Hertz.[100]

In 2010, researchers at the Salk Institute found that synchronous cortical input from the thalamus occurs when there is a salient event in the sensory environment, such as the entrance of a moving object into the receptive field, and they determined the number of synchronous thalamic spikes needed to reliably report a major sensory event to the cortical neurons. The timing of spikes appears to represent analogs, as for example, the object's color or orientation. For example, a spike representing "pinkish red" fires in synchrony with a spike for "round contour" enabling the visual cortex to merge these signals into the recognizable image of a flower pot. [101] By establishing the correct protocols and data representations, scientists will eventually develop the communication packets needed to communicate with the brain from a digital computer. This may occur through specially designed interfaces that communicate through the optic nerve or directly with the cortex, which in one embodiment could produce artificial sensations including vision and, in another instance, represent a novel way of inputting information directly into the brain.

As cities grow and technology takes over the world belief and imagination fade away and so do we.— Julie Kagawa, *The Iron King*

IMPLANTABLE PUMPS

In the future, our brains and central nervous system will have roving bots traveling across our neuronal landscape to target what ails us—such as pain and disease —,or to help amplify our intellect and rate of communication. However, this degree of medical progress will require the confluence of nano-technology and pharmacology. Until then, medical science will continue to make improvements along conventional lines of technology, and every so often move to the next paradigm in a dramatic fashion—a "flash of genius."

Many medical conditions require treatments involving the infusion of a drug for a lengthy period—often for one's entire life. Implantable pumps can meter pain and spasticity from such diseases as cerebral palsy, multiple sclerosis, or brain and spinal cord injury. They also deliver insulin, proteins, glucose, and ions (e.g. sodium, calcium, and potassium) throughout the anatomy and especially to breach the blood-brain barrier, which prevents many drugs from reaching the central nervous system. Some of these pumps are installed in the body, and others are worn externally through a connection to the body through a tiny needle, a pic, or catheter; and in addition a wire that connects to a sensor.

Pumps, in combination with internal catheters, can penetrate blood-brain barrier membranes and infuse drugs directly to the targeted receptors thereby avoiding the undesirable side-effects of orally administered drugs, which may travel to unintended organs. A common application for a "pain pump" is to administer drugs to the fluid-filled subarachnoid or intrathecal space where cerebrospinal fluid bathes and protects the brain and spinal cord. In the intrathecal drug delivery system, a pump the size of a hockey puck is surgically inserted beneath the skin of the abdomen where it delivers medication through a catheter. Because the medication is delivered directly to the spinal cord, symptoms can be controlled with smaller doses than with oral medication (e.g., 1/300 the amount of morphine or baclofen with a pump than when taken orally).

Beta cells within the pancreas make the hormone insulin. With each meal, beta cells release insulin to help the body use and store blood glucose from food. Patients with diabetes may have to inject themselves with synthetic insulin. In type 1 diabetes, the pancreas no longer makes insulin because beta cells have been destroyed. This generally requires the patient to take insulin to absorb glucose from meals. People with type 2 diabetes make insulin, but these patients do not respond well and need diabetes pills or insulin to use glucose for energy. Insulin cannot be taken as a pill because it is broken down during digestion and must be injected to reach the blood stream. One especially frequent use of the implantable pump is to deliver insulin when needed.

One type of insulin pump uses wireless sensors that detect blood sugar levels and communicate the data to a screen, where the patient monitors the readings and injects insulin as needed. Another insulin pumping device, (the size of a small cell phone) is worn externally (often on a bicep) and when the patient pushes a small "on demand button" it delivers a precise dose of rapid-acting synthetic insulin to match the food taken in.

Often patients needing a new heart cannot wait for an available transplant. At least five million Americans suffer some form of heart failure, many requiring a heart transplant, but only about two thousand hearts become available per year. The supply and demand curve drives science to find an artificial heart pump. However, designing a pump from metal and plastic that can keep up with the thirty-five million beats a year has proved a daunting engineering feat, and the expected lifetime of these pump products is limited to a maximum of about 18 months. But this is about to change.

Part of the difference between supply and demand is bridged using a ventricular assist device (VAD) such as the Thoratec Heartmate II, which former Vice President Cheney had installed to keep blood in circulation.[102] VAD technology can be divided into two categories— pulsatile pumps, that mimic the natural pulsing action of the heart, and continuous flow pumps. The latter would not have to pump thirty-five million times a year, and, therefore the expectation is that its life cycle would be considerably extended.

In 2011, doctors at the Texas Heart Institute successfully implanted the first continuous-flow artificial heart using parts from existing ventricular-assist device technology.[103] This artificial heart does not imitate the cardiac muscle pumping action, but uses a pair of pulseless pumps that, turbine-like spin propellers that push blood through the body at a steady rate.

If I have seen further it is by standing on the shoulders of giants.—
Isaac Newton

REAL-TIME IMAGING ANATOMY

Telemedicine currently provides medical care to patients at locations remote from medical facilities. This initiative has been in development for many years and is slowly evolving as an efficient means of treating patients where doctors are unavailable. However, none of these take advantage of in-the-body devices but operate to transmit signals that operate external devices such as EKGs. And none of these are designed to communicate images of a patient's vital organs and bones; for that, patients must to be examined in a medical facility. At present, no portable ultrasonic devices exist that can be applied to the anatomy by relatively unskilled caregivers or emergency personnel and uploaded electronically to be evaluated by a physician at a remote location.

Today, to examine a patient, a skilled medical technician needs to move an ultrasonic device over the affected area of the anatomy while looking at a display to determine if the sonogram or ultrasonic image (both have essentially the same meaning) is being acquired satisfactorily. In the future, however, ultrasonic imaging devices will be self-scanning transmitting anatomical data to a remote location where medical personnel can view the information.

Current technology can scan in elevation (deep) as well as azimuth (side-to-side), to provide a three dimensional data cube of the anatomy. This data cube can be processed using software to produce a variety of image formats, such as conventional planar images, planar images in different scan planes, as well as surface renderings and orthographic presentations. Using technology borrowed from phased array radar, ultrasonic transducers are capable of moving the sonic beams without manually moving an acoustic probe or using a mechanical scanning device. Essentially the beam can be electronically steered, reducing the requirement for the ultrasonic device to be moved manually over the body. The ultrasonic array then would remain stationary, woven beneath the epidermis of the body or applied on top of the body. In either case, in the future, the entire volume of the body organ will be scanned and the data communicated via WiFi or a phone link to a medical facility.

There is nothing remarkable about being immortal; with the exception of mankind, all creatures are immortal, for they know nothing of death. What is divine, terrible, and incomprehensible is to know oneself immortal.—Jorge Luis Borges, *The Immortal*

BIT OF BIOLOGY

Every living thing has a genome that exists both as a material object and as a blueprint with instructions embedded in a chemical alphabet that spells out the architecture for constructing living tissue. In the middle part of the nineteenth century, Gregor Mendel, a clergyman trained as a scientist, began to explore the notion of genetic information (although he did not call it that), through the hybridization of plants having different appearances.[104] He obtained pure breeding pea plants (referred to as the P, for parental,) generation to examine seven traits he was investigating—such as color (some had white flowers, some had purple), length of stem, and overall size. He cross-pollinated two varieties from the P generation that exhibited contrasting phenotypes (e.g., coloration or size) and created offspring referred to as the F1 (filial 1) generation, which in turn were allowed to self-pollinate and create a F2 generation. He found that the F1 generation always had offspring displaying only one of the parental traits (some plants grew tall, others were short). The F2 generation displayed both parent traits. Mendel called the F1 generation trait the dominant trait, which subsequently was also the most frequently observed phenotype in the F2 generation. Reporting on his finding eleven years later, Mendel launched the modern idea of genetics.[105]

In 1909, scientists knew that a subset class of carbohydrates came in several different basic forms: fructose, glucose, and lactose, all of which we know as sugar. In that year, Phoebus Levene discovered that the sugar ribose, found in some nucleic acids, was also present in the nucleus of cells. We now call that RNA. However, a generation would pass before its significance became apparent. In 1926, Thomas Morgan, author of *The Theory of the Gene*, proved that genes were carried on

73

chromosomes, by identifying genes on the chromosomes of the fruit fly.[106] In that same year, Levene found that certain ribose-like molecules were missing one oxygen atom. These nucleic acids are now known as deoxyribonucleic acids, or DNA, which self-organize into structures called chromosomes. Chromosomes encode information for creating and sustaining all life. The sequence of these codes provides the genetic rules for turning on enzymes, manufacturing proteins, tissues, and organs, and ultimately passing on both visible and hidden features. Fast forward 26 years and Watson and Crick solve a major piece of the DNA puzzle.

As part of discovering the double helix structure of DNA, Watson and Crick found that the two threads were comprised of four nucleotide bases, each designated with a letter of the alphabet: adenine (A), guanine (G), thymine (T), and cytosine (C). These wrap around each other to resemble a twisted ladder, the sides of which are a sugar backbone (deoxyribose) and phosphate molecules (pyrimidine and purine), connected by nitrogen-containing chemicals called bases. Each strand comprises a linear arrangement of repeating similar units called nucleotides, which are composed of one sugar, one phosphate, and a nitrogenous base.

Scientists knew that the A and G in purines and the T and C in pyrimidines appeared in the nucleic acid in equal quantities. With this information, Watson and Crick hypothesized that the four-nucleotide bases fit together in specific combinations where A binds to T and G binds to C. For example, one strand of base materials ATCTGGCTA, results in the complementary strand of TAGACCGAT.

...ATCTGGCTA...

...TAGACCGAT...

For a gene sequence to produce a protein molecule, a copy of the gene is transcribed (recorded) as a molecule of ribonucleic acid or RNA. In this way, the DNA plays a role in the initial creation of the blueprint for RNA, which then carries this copy into the deeper regions of the cell where proteins are synthesized. RNA molecules therefore closely relate to the DNA and act as intermediaries between the DNA and the protein or enzyme eventually produced. The range of gene content within the chromosome that transfers to RNA varies, for example, from 2,968 genes on chromosome 1 to 231on the Y chromosome. The gene sizes vary greatly, with the largest known human gene being dystrophin at 2.4 million base pairs. Scientists now estimate the total number of genes at 20—25,000, with nearly 99.9% of the nucleotide bases exactly the same in all people.[107]

However, fewer than two to ten percent of the genome codes for proteins with large repeating sequences (intron sequences) do not code for proteins. Stretches of up to 30,000 cyclical C and G bases occur near gene-rich areas, apparently forming a barrier between genes that perform some important function. For example, the CpG islands are believed to help regulate gene activity. These sequences lead to the type of genetic information that is valuable for gene therapy and drug research. As such they have become a goldmine of numerous institutions, because these novel, isolated, and purified gene sequences are patentable.

In the early 1990s, a consortium of academics from the U.K., the U.S., and Japan set out to decipher the human genome. The hundreds of scientists concentrated on chromosome 22, which contains an estimated 43 million units of DNA. By the end of the decade all the major regions of the chromosome 22 had been cataloged and posted on the Internet, a communications medium in its infancy when the genome project began. The Human Genome Project publishes the genome to make the sequences publicly available. Another group, SNP Consortium also publishes on the Internet single-nucleotide polymorphisms, with a view towards preventing their patenting.[108] We will consider the possible consequences of patenting later on.

However far modern science and technics have fallen short of their inherent possibilities, they have taught mankind at least one lesson: Nothing is impossible.—Lewis Mumford, *Technics and Civilization.*

NANOTECHNOLOGY

The modern biomedical implantable device had its practical beginnings in the 1950s with the invention of the pacemaker. Prior to that, various control devices were developed, but the apparatuses depended on electron tube technology and required dangerously high voltages. They were also simply too large to operate within the body. Implants were only made feasible with the invention of the transistor during the mid '50s, with its small size (size of an aspirin) and low power requirements that made the biomedical implantable device feasible. Today an implantable package that once contained a single transistor now contains millions. However, there remains a vast uncharted territory where even today's micro-technology is too large to fit into.

In 1959, Nobel prize-winning quantum physicist Richard Feynman famously said, "there's plenty of room at the bottom", but it took another twenty-two years and a scanning electron microscope before anyone actually saw what a single atom looks like, and for all practical purposes, the bottom that Feynman referred to. Today, scientists and engineers routinely work with materials smaller than a wavelength of light, using an array of electron microscopes to "see" and to manipulate the last trace of nature that maintains its concreteness. For once inside the atom (atoms are between .062 nm and .520 nm in diameter), matter takes on the characteristics of amorphous energy waves (the subject of particle and quantum physics), somewhat analogous to the transition of an ice cube to indiscernible steam.

How small is a nanometer? A nanometer is one billionth of a meter. The National Nanotechnology Initiative defines nanotechnology as physical bodies existing in the one to 100 nm range. To put this into perspective a nanometer is from 5 to 10 atoms wide. So, a hydrogen atom (0.1 nm wide), lined up into a row of 10 hydrogen atoms, would be the beginning point of what is considered nanotechnology. We can only provide comparisons. The wavelength of green light is 500 nm, which is too large to reveal the breadth of something 50 to 500 times smaller.

Nanotechnology, a word derived from the Greek word *"nano"*, meaning dwarf, applies principals of both physical and biological sciences at a molecular or submicron level. Because of nanotechnology's size, it allows for the manufacture of devices that are orders of magnitude smaller than most biological systems, such as the human cell or its constituents. One DNA nucleotide runs approximately 0.33 nm long.[109] The entire DNA chain contains millions of nucleotides. For example, chromosome 1 measures approximately 220 million base pairs in length.[110] The promise of the nano-device is that it can work its way into the realm of the nucleotide to advance human capability, maintain our bodies, and cure illnesses.

What makes nanotechnology revolutionary is that for the first time, engineers are able to assemble molecules. This leads, atom-by-atom, to more robust architectures. In practice, the building of atom-upon-atom will take place using replicators, devices capable of growing themselves into usable supra-molecular structures, complexes or composites—much as a buckyball, diamond, or snowflake might.[111] In the future these replicated artifacts will be manifested in such things as sensors and logical arrays for measuring physiological activity or can carry out therapies autonomously. Some nanotechnology devices may be driven by chemical reactions, while others are driven through mechanical-like interventions. The front-end work of assembling structures might be micro electromechanical systems (MEMs) used along with sensors to measure conditions within the metabolic pathways, that is, the series of chemical reactions occurring within a cell. We predict these can be used to detect cancer, tuberculosis, and can provide more efficient vaccines, as we anticipate that nano particles can be targeted to specific organ sites.

In other predictions, nano-repairs re-grow tissue, and grow artificial prosthetics. Tissue engineering makes use of artificially stimulated cell proliferation by using suitable nanomaterial-based scaffolding and growth factors. For example, bones can be regrown on carbon nanotube scaffolds. In the long term, experts predict that nano devices will be distributed throughout the brain where they might clone thoughts and then move them into permanent storage, such as external silicon memories, for future retrieval.[112]

The word, nanobot, appears throughout discussion of nanotechnology, often in connection with the notion of replicator. Some within the scientific community envision a time when nanobots, small machine-like artifacts, will responsive to a program, assemble other molecular machine-like artifacts either resembling themselves, or other artifacts. The analogy is the present manufacture of a plastic or metal machine part, using computer-aided-design (CAD) programs that are then feed into computer-aided manufacturing (CAM) machines, such as a vertical milling machine or turret lathe. Another analogy is the robotic assembly worker that manufactures automobiles on an assembly line. However, at the scale of the atom or even the small hundred-atom-wide molecule, the forces of nature, insignificant at the macro level, present obstacles that would need to be understood and dealt with before any viable nanobot replicator were feasible.

In the 2001-2003 timeframe, K. Eric Drexler, founder of the field of molecular nanotechnology, and Richard Smalley, a recipient of the 1996 Nobel Prize in Chemistry for the discovery of the nanomaterial buckminsterfullerene, debated the feasibility of molecular machines which could robotically assemble molecular materials and devices by manipulating individual atoms or molecules. Smalley's position was that van der Waal forces (one of the four fundamental forces in the Universe but rarely of any significance at the scale of human activity), affect the affinity between atoms making them impossible to simply join at will. He asserts that in practice:

Near the center of the typical chemical reaction, the particular atoms that are going to form the new bonds are not the only ones that jiggle around: so do all the atoms they are connected to and the ones connected to these in turn. All these atoms must move in a precise way to ensure that the result of the reaction is the one intended. In an ordinary chemical reaction five to 15 atoms near the reaction site engage in an intricate three-dimensional waltz that is carried out in a cramped region of space measuring no more than a nanometer on each side.[113]

According to Smalley the nano-fingers required to assemble individual atoms would have a certain irreducible size so there would not be sufficient space in the nanometer-size reaction region to accommodate all the fingers necessary to have complete control of the chemistry.

Drexler's response was that his concept dealt with manipulating reactive molecules not individual atoms. The debate did not stop there, as Smalley argued that there would have to be more fingers and larger fingers to deal with the larger molecules, thereby reducing the finesse required in assembling the larger molecules. Drexler noted, first that an experimental result using a scanning tunneling microscope tip would be imaginable, and second that a ribosome, used by DNA in the production of proteins, serves as an example of a natural molecular machine.

Clearly, nature had long ago solved the problem that Smalley and Drexler debated: it continuously replicates natural molecular machines, proteins and complex cells in uncountable numbers. In support of Drexler's position regarding artificial replication, we need only turn to recent advances in synthetic biology and the self-assembly of products in the semiconductor industry (such as quenching, crystallization, polymerization, vapor deposition, solidification, etc.).[114] So, it is possible that that progress will be made towards self-replicating nanobots as molecular assemblers and nano-robots (capable of manipulating individual atoms), build desired products. In support of the theory of self-replicating machines, below, we will discuss von Neumann's universal constructor as a prototype for a replicating process operating in a cellular automata environment.

A cell generally consists of an outer permeable plasma membrane, about 10nm thick, which contains the cytoplasm—the place where the cell's energy apparatus resides. As a significant organelle of the cell, the plasma membrane functions to maintain a stable state, which we refer to as homeostasis. The plasma membrane contains phospholipids, which are complex compounds containing hydrophilic, phosphate group heads and hydrophobic tails of long chain hydrocarbons, called lipids, which dissolve fatty acids. Within the cytoplasm, the nucleus of the cell resides, enclosed by a nuclear membrane, and enshrining the DNA within the structure of its 46 chromosomes. The DNA, in combination with synthesized molecular machines called ribosomals RNA, manufacture proteins and perform functions on biological apparatuses that are less than 25 nm. The plasma membrane communicates between the outer and inner environment through molecular diffusion, moving molecules of protein and oxygen as well as removing the spent products of metabolism.

When something such as a protein is assembled, it is typically hundreds of nanometers in size. Small bacteria are 1000 nm and viruses are 30 to 400 nm. Some cell replicating components, such as the nucleate bases for RNA and DNA, are between one and two nm in size. However the molecular machines that drive these functions are larger; mitochondria for example are between 10,000 and 100,000 nm in diameter. White blood cells are 20,000 nm too large to be within the nanoscale range, but it is within such cells that the nanotechnology does its work when applied directly toward diagnosis and therapy or, perhaps, the manufacture of a specific protein. Future in-the-body nano-machinery will consist of labs-on-a-chip for diagnosis, carbon-based buckyballs carrying cancer fighting drugs, and quantum dots for lighting up diseased cells and as acting as contrast agents for cell-specific non-radioactive imaging.

Proteins are molecules that include enzymes, structural materials, and hormones all of which produce a wide variety of biochemical reactions. All proteins consist of large polymeric molecules, which have chains of smaller amino acids that are held together by covalent bonds. Peptide bonds link amino acids together as chains referred to as polypeptides. Polypeptide sequences form a virtually infinite number of protein structures. These tangled threads can be isolated and synthesized by modern chemical production techniques, making them extremely valuable to the pharmaceutical industry. There will be advantages in nano-technology mechanisms that can produce engineered proteins, *in vivo,* (within the living organism), for warding off viruses and other metabolic dysfunction.

In 2012, UCLA biochemists reportedly designed specialized proteins that assembled themselves to form tiny molecular cages hundreds of times smaller than a single cell. They claimed that these miniature structures would advance methods of drug delivery or even methods of designing artificial vaccines. "This is the first decisive demonstration of an approach that can be used to combine protein molecules together to create a whole array of nanoscale materials," said Todd Yeates, UCLA Professor of Chemistry and Biochemistry and a member of the California NanoSystems Institute at UCLA.[115]

Synthetic nanoparticles used for medical purposes are generally solid, colloidal particles in the form of macromolecular substances that vary in size from 10 to 1000 nanometers. Functionalities can be added to these particles by interfacing them with biological molecules or structures. For example, drugs can be dissolved, entrapped, or absorbed by the nano-macromolecules and form a nanoparticle matrix of other like molecules. Thus far, the integration of nano-materials with biology has led to the development of diagnostic devices, contrast agents, analytical tools, physical therapy applications, and active release drug delivery vehicles, to which they are well-suited because their mechanisms are controllable and their structures smaller than most biological cells. These products will lead to semi-permanent implantable delivery systems, which are preferable to injectable or orally administered drugs, which frequently display first-order kinetics (the blood concentration goes up rapidly, but drops exponentially over time), causing difficulties with toxicity and efficacy, as drug concentration falls below the optimum range. Another application is to replace the current regime of cytotoxic cancer drug delivery, which kills malignant cells, but often damages normal cells. A nanotechnology drug delivery system will target the drug specifically at the malignant tumor, eliminating any side effects to the normal cell.

Other diagnostic uses will come in the form of magnetic nanoparticles, bound to a suitable antibody, to label specific molecules, structures, or microorganisms. Gold nanoparticles will be tagged with short segments of DNA to detect genetic sequences. Nanopore technology for analysis of nucleic acids will convert strings of nucleotides directly into electronic signatures.

Cells contain what is analogous to interconnecting molecular machines or biological motors, assisting in protein synthesis, ADP phosphorylation, and DNA replication. All biological motors have three common elements: a mitotic spindle; a catalytic reactor to supply fuel; and a "power stroke" for fusing or fracturing molecular bonds. Nanotechnology motors have been created that essentially replicate all three of these features. A group at Cornell University has a created a molecular motor in which nanowires act as propellers. The energy is supplied by the body's natural production of ATP (adenosine triphosphate), which otherwise maintains cell structure and serves as one of the main energy sources for human metabolic process essential for locomotion and respiration. Penn State and Rice University collaborated on the development of a nano-molecule with two disk-shaped segments, a top and bottom, joined by a single metal atom (resembling an axle with two wheels). The device is chemically similar to the molecules found in certain hemoglobin (the proteins that carry oxygen in blood). The disks can be joined to construct a double-decker molecule. Fixed in place, the bottom disk serves as a stationary "stator" component that functions as a pivot and the top disk serves as a revolving rotor which turns an "axle." The second nano molecule produces a nanoscale "car" that essentially "drives" on buckminsterfullerene wheels with a three-nanometer wheel-base sized on the order of a DNA cross section. The car can move back and forth in the direction perpendicular to the axles.

Nanotechnology has opened ways to incorporate nanoelectronic devices into the biological process using cellular neural/nonlinear network-based systems. Implementing nanoscale sensors on a cellular neural/nonlinear network-based platform will allow feedback control systems, a basic component in control theory and control mechanisms.

In addition to having developed language, man has also developed means of making, on clay tablets, bits of wood or stone, skins of animals, and paper, more or less permanent marks and scratches which *stand for* language.—S.I. Hayakawa, *Language in Action*

LANGUAGE AT THE BOTTOM

During the last half-century, scientists discovered natural molecular languages inherent in biological processes. At the same time, new computer programming languages were being invented to deal with the complexity of the digital age. This century will see the fusion of computer based and molecular based languages to create new artifacts: part-computer—part biology.

Aided by computers, the sciences solve problems by utilizing processes that manipulate symbols and syntax to achieve levels of semantic interpretation. These are programming languages with built-in intentionality. They reflect the designer's intention, which, when incorporated into in-the-body devices, expresses what software engineers consider "best" or "optimum" to achieve the function he or she is charged with providing. We depend on the designer's ethical duty to only engineer those functions that are essential to the medical prescription. This would imply that a patient's autonomy and beneficence must be protected. The degree that these ethical imperatives are instantiated in the software/hardware products of the future remains to be seen.

However, the assumption is that the in-the-body implants will be for some "medical," reason. We must also not neglect to examine the non-medical uses to which in-the-body technology is employed, and raise the same concerns for observing a person's autonomy and beneficence. In the future products will be embedded for identification, for replacing the features now included in smart cards, and for assisting in outer world communications between ourselves and a range of commercial, governmental, and medical institutions.

Biologist Sir Julian Sorell Huxley, who led the twentieth century in the development of evolutionary synthesis, once wrote:[116]

The critical point in the evolution of man... was when he acquired the use of [language]... the transmission of organized experience by way of tradition, which... largely overrides the automatic process of natural selection as the agent of change.[117]

When language conflates with technology, the two become one and the same, such as when software (in all respects language) driven processes instantiate the hardware that controls vital bodily processes. Passing unnoticed, the selfsame process of expression becomes the technological end-in-itself. Jean-François Lyotard said,

... for the last forty years the 'leading' sciences and technologies have had to do with language: phonology and theories of linguistics, problems of communication and cybernetics, modern theories of algebra and informatics, computers and their languages, problems of translation and the search for areas of compatibility among computer languages, problems of information storage and data banks, telematics and the perfection of intelligent terminals, to paradoxology...[118]

All living things utilize one or another form of communication (utterances, color, movement, pheromones).[119] Creatures also communicate through biological signaling (such as that which exists within the gene or in one's ideas), of information to keep them united, and to pass on the knowhow to build the artifacts necessary to ensure a future. The human genome expresses a physical form, but also comprises a form that transmits information to assure a future as it replicates. Each gene iterates upon the four nucleic acids to form molecules of larger and more complex statements of form.

Thomson, the scientist who patented the method to produce stem cells from human embryos, observed, "A stem cell replaces itself through proliferation for prolonged periods (self-renewal) and gives rise to one or more differentiated cell types." Yet, how does it know what to replicate? On the molecular level, elements from a finite group of atoms join together based upon affinities caused by electrons in various outer atomic orbits or shells. These affinities attract certain other elements, which assemble themselves into larger molecules. We might consider that these molecules comprise the "letters" in what become macro molecules having analogous sentences, syntax that form of living organisms. These "languages" occur in nature and communicating at the cell-level about how to construct themselves and how to survive. The manner of biological communication varies and, it may be deemed chemical or electrical, but in reality, we refer to one or the other depending on our point of view. The RNA and DNA molecule suggests the most profound "micro-sentences" known to science. For in these molecules we find contained what persists as life on our planet, a form of communication that transmits knowhow. At the macro level of communication, we have the fully formed organism, whether a single celled amoebae or the 50 trillion celled *Homo sapiens*. At the level of the species, the means of communicating takes various forms: tactile, visual, vocal-auditory, chemical or even electrical. The simpler the organism, the simpler the code or "vocabulary"— fewer sentences make up its language.

Men have become the tools of their tools—Henry Thoreau

A GIZMO IS A GIZMO BY ANY NAME

Interest in electricity as a newfound phenomenon had been building for nearly a hundred years by the time it was well enough understood to produce electromagnets, motors and generators. In 1840, Samuel Morse had the idea that he could interrupt electric current passing through an electromagnet causing an apparatus to perforate paper according to a decipherable coded message.[120] Hammering out a code using paper perforations was not completely unknown in Morse's time, but his widely adopted method lead to the first nationwide electrically-based communication system.[121]

Following Morse's telegraph, Bell's telephone sparked a world of voice communication and Marconi's and Armstrong's radios each led to the creation of massive networks that essentially shrank the globe. These inventions required an enormous public commitment to creating an electrical infrastructure that included transmitting stations, electric wire manufacturers, electricity generating stations, and of course capital to build factories, and educational institutions. This same phenomenon was to repeat itself when ARPAnet established intercommunicating computers that eventually lead to the Internet. What ARPAnet did for the Internet, medical electronics will do for the anatomy, tying internal medical devices into a colossal in-the-body communications network.

In the early nineteenth century Charles Babbage and his protégé Augusta Ada Byron Lovelace (the daughter of the English poet Lord Byron) developed a calculating engine made from pulleys, levers, and brass, incorporating the logic of George Boole, the kind of zero-one logic that serves as a foundation for today's computer. But it was not until the early 1940s that Howard Aiken, a Harvard University mathematician, created the first working digital computer along the lines of the Babbage computer. Unlike mechanical computing, this machine used vacuum tubes to switch circuits on and off— a fundamental requirement of all digital computers. The computer also received instructions fed from a roll of punched paper tape or from Hollerith coded punched cards (once ubiquitous IBM cards).[122] A demarcation was drawn between the hardware of the computational device from the software (the data, the programming rules or the instructions).[123]

In this same era, mathematician Alan Turing theorized how a computer might simulate thinking by reading instructions and altering its process based upon what it read.[124] The idea analogized how humans might experience and then change their behavior or thought pattern. We will see later how these same concepts of experience and behavior change surround the DNA code and its process for manufacturing proteins.

What is regarded as computer hardware, microprocessors, switches, display technology, printer mechanisms provide the apparatuses that operate under the control of software. Software comprises the states that a computer memory adopts, based upon electrical signals generated in response to a program, a set of instructions authored by a programmer. With this summary explanation much is missing, but this is the general idea. Early in the development of computers, engineers determined that for the types of computation that the computer was typically used for, architecture should maintain a separation between the parts of the computer that serves to process the data from the data itself.

However, there are other types of computers, too, some that do not perform calculations based on conventional arithmetic rules and others that do not employ digital one-zero logic. Before the digital computer, there were analog computers—the Babbage and the Hollerith-type tabulation machines were early examples. Later examples were analog ballistic computers used in ship and aircraft fire control systems. Countless analog computers are found in every day devices, although not much attention is drawn to them. Other computers, such as optical computers, do not employ microprocessors as such, but use lasers, lens, mirrors and holograms to process complex wave fronts, such as produced by light and sound, to rapidly recognize patterns. Quantum computers use quantum mechanics to perform operations on data by encoding them as quantum properties. Unlike the digital computer, the answers are not absolute, but probabilistic outcomes.

Different kinds of computers direct our attention to their physical embodiment, materials, mechanisms and architecture, required to fulfill an intended application. Over time these might change, so a Babbage computer made from metal, pulleys and a metal framework, has in the modern era been replaced by silicon, transistors and a plastic or polymer framework. The physical embodiment of the modern microprocessor, essentially the modern digital computer, will in the future be embodied in the artifacts of molecular computers using combinations of organic and inorganic materials embodied in nano technology. But, the technology itself is only part of the equation. The other part concerns the programming, that piece that along with smaller, faster computers, adds that component we associate with knowledge or intelligence.

The IBM Watson computer was developed to run artificial intelligence programs of the kind that rapidly search and spew forth answers to unstructured questions. In 2011, its power was demonstrated on *Jeopardy*, before millions of viewers, when it outscored the show's all-time money winner Brad Rutter and the record holder for the longest show's championship streak Ken Jennings.

The machine is capable of processing 80 trillion operations (teraflops) per second. It runs about 2,800 processor cores and has 16 terabytes of working memory... To build a body of knowledge for Watson, the researchers amassed 200 million pages of content, both structured and unstructured, across 4 terabytes of disks. It searches for matches and then uses about 6 million logic rules to determine the best answers.[125]

It is not a Watson-like computer in its present form factor, that is size, architecture or power requirements, or made from the same materials, which is relevant to our anatomical future, but what happens when materials change so that the computer is compatible with the human form, and its computational power packs into the volume approximating a sphere the size of a red blood cell. Will we then witness emergent qualities, those that cannot be predicted from the intended design or program itself?

Artificial intelligence undoubtedly will be built into in-the-body processors, which will exhibit self-learning, self-replication, regulation, and control. Each of these processes will be achieved by different architectures, executing programs, some of which may be computational and others applied, not to number crunching, but as constructors of biological entities. First generation devices will interpret data, self-learn, regulate, and control physiological functions, but will operate much as conventional computers do. These will be specialized to monitor sensors, provide electrical stimulation to anatomical sites, and communicate with the outer world. Second generation devices with new architectures will interface directly with the genome in the production of engineered, synthetic genes, for therapy and enhanced physiologies.

Computers in combination with biological function at the cell level will be designed to reduce illness, enrich various phenotypic features, such as intelligence or resilience and extend life. Ends possible, on a specific level are only speculative, and, therefore the potential for so-called side effects or unintended consequences are also speculative. What is not speculative is that as the population ages, there will be a proliferation of in-the-body computers with more and more sophisticated programming, much of it embodying artificial intelligence for the applications previously mentioned.

In addition to the main objective of keeping us healthy and enhanced, ancillary programs will be likely to modify and update the typical operating systems and application programs for purposes that do not necessarily consider our individual well-being. Computer systems, for instance, may be programmed to assist manufacturers, governments, and the medical community to improve products, track population illnesses, and facilitate such matters as record retrieval. It is likely that periodic updates would be downloaded from a computing "cloud," which could potentially carry errant code, unintended programming errors, or intended programmed viruses. Much of the programming will involve forms of artificial intelligence, since by definition: some of the subsystems will be installed for purposes of recognizing patterns, detecting trends, building upon what it already "knows," and melding it with data from outside-the-body. Let's briefly explore the realm of artificial intelligence to speculate on what is on the horizon.

Computing technology is racing in five significant directions: greater complexity in increasingly smaller volumes, increasingly faster processing speeds, greater amounts of information storage, increasing bandwidth and channel capacities, more intelligence (albeit artificial), and all of the above merging with the human anatomy.

For the next two decades, computer processors will continue to shrink to the size of a bacteria, 2 microns or micrometers (μm) long and 0.5 μm in diameter, with a cell volume of 0.6 - 0.7 μm^3, while increasing their processing speeds to upwards of 10^{16} (10 trillion) operations per second, referred to as flops.[126] This small size will allow computers to live both within and alongside the human cell. The level of 10 trillion flops exceeds the processing rate, at roughly 10^{15} FPS, of the human brain. This will allow adjuncts to memory to coexist within us, providing Watson-like artificial intelligence to be accessible as desired. Some computers will control microelectromechanical systems (MEMS) such as nano-sized motors that will range in size from 10^{-9} (a few hundred atoms across) to 10^{-5}—the diameter of a white blood cell. These traveling nano-bots will course through arteries and cellular membranes to deliver drugs and destroy pathogens. Many of these technologies will employ cell-specific targeted nanoparticles laced with gold or other metals that will improve imagining radiation diagnostics and therapy. The noted physicist Michio Kaku writes:

... it's still premature to say where nanotechnology will go. However, one place where technology may go is inside our body... these atomic machines will be the size of blood cells and perhaps they would be able to perform useful functions like regulating and sensing our health, and perhaps zapping cancer cells and viruses in the process. [127]

That common seed between artificial systems and biology of organization and pattern has recently emerged in the form of the molecular computer—a device capable of solving a wide category of mathematical problems. This invention is remarkable because it is neither electronic nor mechanical, but is richly informational DNA, employing a combination-seeking apparatus fundamental to its molecular makeup. Those mathematical problems looking for solutions in combinatorial logic may have found a device well equipped to serve its needs. A molecular computation using DNA was first reported in 1994 by Leonard Adleman, at the University of California. In what would seem the first practical demonstration of the DNA molecule as a computational element, he solved a classical mathematical problem referred to as the "Traveling Salesman" or Hamiltonian Path Problem.[128] Although a pencil and paper solution to the problem teases the most astute mathematician, a statement of the problem does not: What route does a salesman take to get from one island to another crossing over bridges connecting each to the other, without crossing the same bridge twice? Each point of departure from a landmass over a bridge would be termed a vertex. If you imagine six islands, for example, each can be further imagined as a point or vertex on a hexagon. The lines, between the points would then represent the bridges. In his application of the DNA, he would represent the various bridges and islands as strands of DNA. Since Adelman, the technology of molecular computing using synthetic DNA has skyrocketed, and although most applications are still in research laboratories, in the near future a synthetic DNA computer will be coming to research hospitals near you for one or another trial installation in the human anatomy.[129]

Non DNA-type molecular computers will use molecular transistors made from benzenethiol, rotaxane, and graphene chemicals. Recently scientists discovered that benzenethiol can be made to act like a valve, replacing the base or gate element in a transistor, and thereby control the flow of electrons through the device, turning current on and off—the basic requirement for digital computation. Rotaxane molecules can rotate around the axis of the dumbbell like wheel with an axle or slide its hubs along its axis one side to the other, so that controlling the position the molecule acts as a switch that assumes different states.

Molecular computers would consist, as the current silicon microprocessor of many molecular transistors; however, scientists have not yet solved how to wire them together. When scientists solve this problem, and there is every expectation they will, the microscopic size of molecular computers will allow their insertion into the nucleus of a cell. These will have a highly dense information storage capacity (1 bit per cubic nanometer), rich parallelism (allowing multiple operations to proceed simultaneously), and energy efficiency. No active power sources will be required as they perform 10^{19} operations using one joule of energy (energy released when an apple falls one meter to the ground). These combined features, especially the low energy requirement, places them in situations that will never be realized by the silicon microprocessor. One gram of DNA (approximately 1 cc when dry) can hold as much information as approximately one trillion CDs. The latest supercomputers operate in the range of about 10^{12} or 1 billion operations per second. By using DNA molecules with molecular computers, speeds as high as 10^{15} operations per second is within reach.

Unlike their silicon counterparts, molecular computers will eventually lead to targeted medicine, where they will be combined with nanoparticles to seek out abnormal cells and deliver the nano-doses of drugs, where and when they are needed, thereby avoiding the kinds of system-drug-flooding that current drug therapy causes.

We are quickly reaching a point where the theoretical constructs between conventional computers and molecular computers are disappearing. The theories that hold for the computer in the abstract do not depend on physical implementation. Considerable progress has been made demonstrating that molecular computers follow the same general principles as a Turing machine, the theoretical underpinning for how digital hardware and software combine to carry out the simplest and most complex computations in the history of civilization.[130]

Technology, when misused, poisons air, soil, water and lives. But a world without technology would be prey to something worse: the impersonal ruthlessness of the natural order, in which the health of a species depends on relentless sacrifice of the weak.—*New York Times*, editorial, 29 August 1986.

MERGING STRUCTURE AND PROCESS

The year is 2050 and Eve's son Cable was a promising concert pianist studying music at a prominent university, before he severed the adductor pollicis muscle on his left hand, the one that mainly controls the extension of the thumb. Dr. Newberg a leading hand reconstruction surgeon has recommended a replication procedure, which can regrow the section of the damaged muscle, but without scar tissue. The doctor explains that an elective feature of the surgery is to choose the desired extension of the thumb, effectively increasing the hand span from the small finger to the tip of the thumb. Cable tells the doctor that before his injury he could stretch ten piano keys, referred to as a tenth, but would very much improve his virtuosity if he could expand his reach two keys, to a twelfth. Dr. Newberg believes he can achieve the desired results using a combination product that employs stem cell, synthetic genetic biology and a replicator computer that drives the process. Let us explore how close the Cable's of the world are to this kind of makeover.

The materials for future computers will range from the standard silicon based products of today to molecular and nano-carbon based products of tomorrow. Each technology will target applications where best fits into the incredibly small vacancies nature makes available on the biological scale. And as in the macro world, embedded computers of the future will have specialized uses and as most computers are by definition computational, some will subtract and add or employ standard logical operations in the old fashioned way. Other computers will be designed for not computation, but for replication, for making computers like itself or reproducing artificial biological products. These replication devices will incorporate rules (embodied in their instruction repertoire) that transform one pattern into another pattern, one substance into another substance, one process into another process. For example, imagine that the computer senses a chemical, analyzes it in a small chamber and puts out the same or a different chemical. Or, it senses a molecule and replicates or duplicates itself in the structural twin of that molecule. In this way, it operates as a self-replicating machine. It acts as a constructor of complex molecules, such as proteins or enzymes. Obviously such a device is not the silicon variety, but it is conceivably a molecular or nano-carbon based device that can construct another like device. It is imaginable that a DNA machine can transcribe and assemble DNA base pairs to make protein. And, finally, it may be possible that a compliment of computers can be organized to work in concert to do some combination of all the above.

In the late 1940s the computer scientist von Neumann proposed a computer useful for applications that explored self-replication, in which he envisioned a computer where the mother automaton could construct all portions of daughter automatons.[131] His goal was to:

> ... abstract from the natural biological self-reproduction the logical form of the reproduction process, independent of its material realization in any particular physio-chemical form. He was able to create a universal Turing machine consisting of a two dimensional automaton... Self-reproduction then followed as a special case when the machine described on the constructor's input was the constructor itself.[132]

In 1951, von Neumann then considered that it might be possible to create computer languages that constituted both the program and the data.[133] These were immediately recognized to be self-replicating programs. [134] DNA is itself a self-replicating machine, where the program and the "data" are instantiated in the codons to be transcribed and then translated. In 2011, a team of scientists created a "bent triple helix" structure based around three double

96

helix molecules, each made from a short strand of DNA. By treating each group of three double-helices, as a code letter, in principle, one can construct self-replicating structures that encode information. [135] In fact, along similar lines, more than one mathematician, computer scientist or physicist has considered the amount of information one would need to construct an android capable of replicating itself. [136]

In considering developing self-replicating programming, von Neumann and Stanislow Ulam collaborated on an idea suggesting the state of computer data contains information that determines the next state of the computer. Imagine that these states are not only electrical states, but also potentially physical states. An analogy is a water molecule that at a given temperature crystallizes and forms a branch of a snow flake. That new branch then forms yet another branch and so forth. The scientists' work gave birth to cellular automata. This field focuses on the dynamics of automata that express the rules for self-replicating systems, which also express the rules or language of reproduction. John Casti, a researcher at the Santa Fe Institute, writes:

> While human languages are unimaginably complex, there is another type of language employed by nature that offers many possibilities for analysis using the automata-theoretic ideas... as a first approximation we can consider the DNA molecule as a one-dimensional cellular automaton having for possible values per cell, which we shall label A, C, G, and T, representing the four nucleotide bases at Adenine, Cystosine, Guanine and Thymine... The Central Dogma of Molecular Biology, due to Francis Crick, is a statement about the flow of information in the cell. Compactly, it states that DNA →RNA→ Protein... The Central Dogma is a description of the way in which a DNA sequence specifies both its own copying (replication) in the synthesis of proteins (processes of transcription and translation). Replication is carried out by means of the base-pairing complementarity of the two DNA strands,

with the cellular DNA-synthesizing machinery reading each strand to form its complementary strand. The protein-synthesizing procedure proceeds in two steps in the transcription step, a strand of messenger RNA (mRNA) is formed, again as a complement to one of the DNA strands, but with the base element U (Uracil) taking the place of T (Thymine).[137]

The sequence of DNA nucleotides represents a pattern that influences the development of biological systems. In effect such systems depend analogously on a vocabulary and syntax interpreted according to biological/chemical rules to transform the properties of one substance into another. A chemical process often transforms of two materials into a third material that consists of either the properties or potentialities of the first two products. This occurs in the construction of RNA and the subsequent production of proteins. However, all of these phenomena are forms of information, represented as patterned expressions. We can symbolically represent the manner in which they transform, create rules and in many cases deal with the phenomena artificially as computer programs that model the behavior we might be interested in observing. In this respect, artificial systems qua computers and the natural system comprised of biological material share a common seed, patterned expressions, and depending on the computer the outputs can likewise behave to construct or replicate what the intelligent designer of the program intended.

In a demonstration of a replicator John Conway published a set of rules, dubbed the Game of Life, that involved the creation and annihilation of points that were arranged on a checkerboard-like pattern, what mathematicians refer to as a two dimensional array. [138]The transition rules, applied to a central point, were applied to the existence or nonexistence of the combinations of the eight adjoining neighbors. Conway studied the effect of the following rule: (1) if an empty central space exists whose immediate neighborhood contains exactly three points, the central space would be filled with a new point; and (2) if a central point had in its immediate neighborhood consisting of two or three points, the central point would remain on the board; and (3) if the immediate neighborhood of a central point contained four or more points, the central point would be removed. The rule as applied to each point or element determined whether the element's future either remained in play or was summarily removed. If the process of adding, removing or keeping the element in place at each square is repeated, over and over, you might see that checkerboard has a pattern of checkers different from the starting pattern.

Going back and forth in this manner, and examining the pattern, one or more patterns eventually result. One might observe that the pieces gradually disappear from the checkerboards and therefore any hint of a pattern vanishes. As a next possibility, the pattern might eventually stabilize in one of two ways: (a) it would grow to a fixed size and simple stop there or (b) it would grow to a certain size and then retract in size and then grow again in a repetition of expansion and contraction. Finally, the pieces might exhibit a growth pattern until they were to run out of checkerboard space. Self-contained stable growths do not depend on lattice-like arrays, such as checkerboards, but could be created from any number of conceivable arrangements of points in one, two or in the case of the real world three dimensional space. [139,140]

Without digressing into a treatise on cellular automata, these automata patterns imitate a variety of molecular growth activity, and illustrate the emergence of patterns from memes, temes, languages, (both human and computer programmed) and natural phenomena (as for example protein construction from RNA transcription). Wolfram, says,

> ... there are examples all over physics and biology of systems that look like that, that grow in exactly that way: crystal growth, for example, cell growth in embryos, the organization of cells in the brain, and so on. The important thing is that the mathematical features of cellular automata are the same

mathematical features that are giving rise to complexity in a lot of the world's physical systems.[141]

Cellular automata does not represent anything unless we give meaning to its rules and states as represented by the symbols we choose (points, lines, zeros, ones, or even lights). Until we assign meaning anyone of the several examples of automata amounts to nothing, except to demonstrate that we can create novel patterns and sequences. John Casti writes:

> The crucial point to note about von Neumann solution is the way in which information on the blueprint is used in two fundamentally different ways first, it's treated as a set of instructions to be interpreted which, when executed, cause the construction of a machine somewhere else in the automation array. Thereafter the information is treated as uninterpreted data, which must be copied and attached to the new machine. These two different uses of information are also found in natural reproduction, interpreted instructions corresponding to the process of gene translation, while copying the uninterrupted data correspond to the process of transcription. These are exactly the processes we discussed above in connection with cellular DNA, and it's worth noting that von Neumann came to discover the need for these two different uses of information several years before their discovery by biologists working on the mysteries of DNA. The only difference between the way von Neumann arrange things in the way nature does it is by is that von Neumann arbitrarily chose to have the copying process carried out at the construction phase, whereas nature copies the DNA early on in the cellular processes.[142]

DNA represents a short hand not only for describing the molecular constituents, but for making the directional distinction that nature takes from the genotype to the phenotype. DNA creates the configuration for all subsequent tissue development. For example, cloning begins in the recognition of this fact. Many millions of DNA transactions form proteins, which construct tissue, which make an organism. The manufacture of a protein begins with first making an RNA copy of DNA using a process called transcription. The transcribed RNA copy contains sequences of A, U, C, and G that carry the same information as the sequence of A, T, C, and G in the DNA by forming a complement or cDNA of one of the DNA genes. The RNA molecule, now called messenger mRNA, then moves to a location in the cell where the molecule releases from the DNA molecule.

Actually synthesizing proteins from mRNA involves structures called ribosome that bind to the mRNA. The ribosomes "read" the information in the mRNA, by shifting along the strand of mRNA translating the non-overlapping section of codons three nucleotides at a time, adding the amino acid specified by that codon to a growing polypeptide chain until it reaches a stop codon.[143] If computer processes, using synthetic DNA, are injected into the process of protein production, it may well be in constructing codons that are designed to be inserted into the protein matrix. Scientists as UCLA are advancing protein engineering targeted to better drug delivery methods or artificial vaccines and build nanostructures by assembling designed protein domains in a designed rigid configurations.[144]

One of the projects these new engineers and programmers will engage in concern correcting DNA errors by using stem cells or by inserting DNA, natural or man-made into living cells for therapeutic reasons. And, carrying out these objectives will force the development of new biological computing platforms that rewrite and construct natural systems. In many ways, this is the world engineers have always lived in, except there is a change in materials, from the inorganic stuff of rocks and metals, to the organic world of sugar and acids, from the processors at our fingertips to those that are deep-seated in the cellular terra firma of our inner bodies.

Embryonic stem cells, extracted from human embryos, are the foundation of a new era of regenerative medicine, one in which doctors might be able to use the cells to develop cell types that are of great medical importance to diseases of the liver, muscle, nerve, pancreas, and bone. The technology will also find application in the specific production of insulin-producing cells and nerve cells to treat diabetes and Parkinson's disease, respectively. In general, stem cells are undifferentiated—that means that they have not committed themselves (referred to as pluripotent) to develop in one or another way, until they mature in environments that give rise to advanced functional cells. For example, a hematopoietic stem cell may give rise to any of the five major terminally differentiated white blood cell types. Skin stem cells produce skin cells. The human embryo cell possesses the capability of developing into any organ or tissue type and at least potentially into a complete embryo. In the future stem cell technology will merge with synthetic biology to change the architecture of the use to which the cell is put.

As mentioned earlier, James Thomson derived human embryo cell lines from human blastocysts-stage preimplantation embryos produced by in vitro fertilization. Three years later, he proved to the satisfaction of the patent office that he had developed a novel process of isolating and keeping alive stem cells that were useful in manufacturing a variety of other cells. As such, he secured a patent, which he assigned to the University of Wisconsin. The patent covers both the method of isolating the cells and the cells themselves. Thomson's invention provides a purified preparation of a primate embryonic pluripotent stem cell line. The cell lines: (1) are capable of indefinite proliferation in vitro in an undifferentiated state; (2) are capable of differentiation to derivatives, of all three embryonic germ layers (endoderm, mesoderm, and ectoderm); and (3) maintain a normal karyotype (DNA complex) throughout prolonged culture.

Thomson's technology raises important economic and moral issues that will emerge with the marriage of stem cells and synthetic DNA. Here, science will create new organ designs, designs that will enhance the human body by adding biological extensions, possibly creating vectors for carrying extra chromosomes to a world where lives will carry on into centuries.

The current horizon in synthetic biology stretches from nearly a quarter-century ago when a team at ETH Zurich created DNA consisting of two invented, artificial genetic base pairs, designed to work with the AGTC base pairs that are the foundational elements of nature's genome. In 2000, scientists reported constructing two devices that worked inside the *Escherichia coli* cells by altering its DNA sequences to cause the *E. coli* to blink predictably—that is turn on and turn off. [145] The other device contained a feedback loop allowing it to flip-flop, toggling between two states like a common electronic circuit used in computer memory. Typically such devices, as is the case with these new bioengineered devices, perform logical operations that correspond to Boolean logic,—the arithmetic of computers, which carry out simple operations such as combining the presence of two inputs into one output. In the language of Boolean logic, these are referred to as "AND and "Not AND" or "NAND" functions, which according to mathematical theory, if properly arranged, can carry out any imaginable, realizable mathematical function. This, of course, is at the very foundation of the modern digital computer.

Other scientists also are creating the computational elements required for constructing full scale bio-computers gate by gate, using synthetic DNA technology as the logic device. Chris Voigt, a synthetic biologist at the University of California—San Francisco recently engineered a one bacterial system to regulate gene expression in response to red light and another to sense its environment and conditionally invade cancer cells. As reported in *Nature*:

> ... a bacterial system that is switched between different states by red light. The system consists of a synthetic sensor kinase that allows a lawn of

bacteria to function as a biological film, such that the projection of a pattern of light on to the bacteria produces a high-definition (about 100 megapixels per square inch), two-dimensional chemical image. This spatial control of bacterial gene expression could be used to 'print' complex biological materials, for example, and to investigate signalling pathways through precise spatial and temporal control of their phosphorylation steps.[146]

Voigt and his colleagues also successfully demonstrated a logical AND gate inside bacteria. This later achievement is on a par with the electronic AND gate invention used by electrical engineers in developing control circuits and eventually led to the creation of the electronic computer. These synthetic sensors and logical switches have been used to control the assembly and function of such things as secretion needles that export spider silk proteins or a photosynthetic apparatus responsible for converting light into chemical energy. No one disputes the implications for this technology to improve our health. The unanswered question is: how soon. Great motivation exists to expand biological gate technology, to press forward with nano-processors—the hardware and software of the succeeding generations. Installed into the human anatomy these will carry out objectives of well-being, enhanced existence, improved intelligence, and longevity. Inventors are feverously developing a range of applications for medicine, biofuels, and to arrest global warming. The artifacts and processes serve as prime candidates for developing parallel computers and computers that do not run conventional serial programming agendas, but rather programs based on cellular automata.

For nearly a dozen years, these new synthesized genetic parts have been routinely fabricated and stored as "off the self" items, much like electronic transistors, with M.I.T.'s Registry of Standard Biological Parts. Today the registry contains over 3400 synthetic DNA parts, making its inventory available to academic labs, scientists, and university students who can download software tools through a series of application programming interfaces (APIs).

Synthetic DNA will be used to create new life forms or to reconfigure existing cellular metabolic pathways to perform new functions— such as the manufacture of chemicals and drugs.[147] However, the synthetic biologist's main goal is to build genetic devices as adjuncts to living cells, where cells supply the energy and materials, and the biochemical ecological systems that decode DNA into messenger RNA and finally manufacture protein.[148]

A leading scientist in this burgeoning field is Nobel Laureate Hamilton Smith. In 1995 he helped sequence the first bacterial genome, *Haemophilus influenza*, and later the human genome at Celera Genomics. In 2003, he was part of a group that synthetically assembled the genome of a virus, Phi X 174 bacteriophage. His latest efforts at the J. Craig Venter Institute are to partially synthesize a species of bacterium derived from the genome of *Mycoplasma genitalium*. In 2010, his team announced they had assembled a complete genome of millions of base pairs, inserted it into a cell, and caused the cell to begin replicating.

To create Smith's replicating cell, the DNA code was transcribed as a computer file, edited with new code, sequenced, and put together using yeast and other cells, and finally it was transplanted into a cell from which all genetic material was removed. The cell divided and came under control of the new genome. [149] This was a stunning accomplishment, as it demonstrated proof of concept that "The Central Dogma" can be propagated from a computer file, edited, sequenced, and reassembled—virtually overlapping the boundaries between life and machines— yielding "truly programmable organisms."[150, 151]

Warning: As of this writing, scientists have not installed a computer at the cellular level to construct or alter a gene from synthetic and/or natural DNA or combination of the two. But let us engage in pure speculation for a moment and imagine if a microprocessor constructed either from conventional silicon or perhaps from the newer Voigt-like biological bag of parts, were to carry out the steps to manufacture synthetic DNA, and to install the DNA at sites dictated by a diagnosis and prescription using the "Central Dogma of Molecular Biology". Let's call this process A.I. for artificial intelligence. Modified, an account for the system would appear as:

A.I. →DNA →RNA→ Protein

The A.I. machine would undoubtedly be one where the synthetic DNA molecules were constructed through technology installed into a live DNA genome. Also, we should broaden out our thinking and consider that the DNA model suggested here may not be directly affected at the nucleotide (AGCT) level, but at the epigenetic level where A.I influences the tags, which act as cellular memories and react to signals from inside hormones, enzymes, or other chemicals released via a nano-lab to turn the DNA genes on or off.

Several hundred cells, each manufactured from the DNA blueprint, comprise the inventory of muscle, nerve, bone, and skin, which contribute to the functioning of the component parts present in all mammals. Many millions of cells,—some layered, some solid, some transparent,—form tissue of many varieties. The DNA may supply the information, but two mechanisms are required to transform natural elements and chemicals into the various tissues: a control switch, to initiate certain processes and a chemical reaction. These are the domain of protein production and epigenetics—a mechanism through which the body regulates gene expression.

From the recent successes of the Human Genome Project in deciphering DNA, scientists rapidly shifted to cataloging the library of millions of proteins that account for cell structure and state of biological processes. In the latest addition to the technological lexicon, proteomics now leads the way in the search for the Solomon's mine of therapeutics. DNA may be the fundamental blueprint for life, but the final engineered products configure themselves from massive inventories of distinct proteins. These three-dimensional combinations of molecules fasten together like beams, trusses, and trestles to create both structure and dynamics in the performance of form and function without which life of any species would not exist. It is precisely this that A.I. intends to re-engineer.

Epigenetics is the study of heritable changes in gene expression or cellular phenotype caused by mechanisms other than changes in the underlying DNA sequence. These changes are functionally relevant modifications to the genome that do not involve the ACGT nucleotide sequence, but do serve to regulate gene expression. The DNA in our body is wrapped around proteins called histones. Both the DNA and histones are covered with chemical tags, methyl tags, and acetyl tags. This structure, called the epigenome, wraps tightly around the DNA's inactive genes to make them unreadable, while relaxing the active genes makes them accessible. For example, DNA methylation and histone modification, both of which serve to regulate gene expression, do not alter the underlying DNA sequence. Although the DNA code remains fixed for life, the epigenome is plastic—able to stretch linearly or to wind itself into a coil. When coiled tightly it may reduce the mRNA translation and protein production, and when stretched may increase mRNA translation and protein production. The epigenetic tags act as cellular memories, reacting to signals from inside the cell, neighboring cells, and to hormones released in one or another part of the body. They also react to environmental factors as the situation dictates: illness, habit, such as smoking, diet, and stress that produce hormones or brain chemicals to turn genes on or off. Gene regulatory proteins help recruit enzymes that add methyl tags to the DNA, thereby influencing the mRNA production. The epigenetic tags attached to the DNA strand is replicated in new complementary strand when the DNA divides, so that the tag moves with the DNA, carrying with it the memory of the particular experiences that caused its production in the first place.

The gene regulatory proteins that help recruit these enzymes would be wholly controlled by the A.I. device mentioned earlier. Theoretically, they would affect epigenetic tags carrying with it the memory of the particular A.I. experiences programmed by the intelligent designer.

This would seem to be where this is all heading. The rules of transformation, the A.I. portion, depend on the atomic affinities that certain combinations of codons present to fragments of amino molecular material to build upon. Mutations, or rule changes in a germ line will produce variants on the DNA molecule in subsequent stages of evolution.

Let me summarize: The resemblance among design patterns in DNA, memes, temes, and artificial cellular automata, to the extent that they employ automata-theoretic processes, opens a vast swath of cyborg related possibilities, leading from A.I to DNA, to constructs of a new social reality. Every elemental atom and molecule represents an intrinsic artifact of our universe. As organizations of molecules, biological cells fit within this intrinsic group of things and serve to carry out structural, electrochemical and ultimately, what we generally refer to as biological function. Invented artifacts, such as an A.I machine may be programmed to direct progressive modification of this intrinsic structure. A.I machines may be programmed to operate autonomously, that is without human interference. As such they may become teme-like machines, operating to create other machines. A.I. machine programs, although directly altering DNA structure or epigenetic controls, are not intrinsic to nature. We may, for purposes of analysis, think of these artificial DNA structures as nature's genetic code, but is not nature's code; it is only a product of our imagination. The unalloyed cell, one not modified by and A.I. machine, had an inherent pattern, has objective, observer independent existence, unlike the computer or programming codes to which we assign meaning, or the kinds of codes that we design and which cause a computer to behave in an assigned manner.

A prediction about when cyborg related possibilities, leading from A.I to DNA to repair Eve's son Cable's hand by the year the year 2050 is impossible to make at this time. From the perspective of cellular automata, we know that small changes in rules or starting conditions make large changes in the evolution of games, population growths and cell growth. We know that we can regrow cells based upon stem cell technology. We know that synthetic DNA is being produced. When and exactly how these technologies will merge is yet to be answered, but there should be little doubt that the merger is certain.

There's a battle outside, And it is ragin', It'll soon shake your windows, And rattle your walls For the times they are a-changin'.— Bob Dylan, *The Times They are A-changin'*

TECTONIC FUTURE

Computers, cell phones, the Internet, Google, Facebook, etc. as some believe have changed our brain's biology. Whether true or not, unquestionably the widespread proliferation of in-the-body technologies will result in a change in our social reality: how we interact, how we assess each other's vulnerabilities and capabilities, how we adapt to new demographics based on the haves and have nots of the new technology, how we prepare defenses to oppose those who would hold us hostage to intellectual property monopolies, short supplies, cyber viruses, and even future electronic wars, where radiation could potentially disable our internal machinery.

As we progress further into this century, the integration of technology into the human anatomy, especially RFIDs and more robust microcomputers, will become an increasingly common event, not much different from body piercing or dental implants. And, except for individuals with obvious limb prosthesis, one might not see a clear demarcation between the artificially enhanced or remediated individual any more than one might make a distinction between those with and those without a face lift, dental implant, or hair transplant. Consequently, out of sight may mean that government and private institutions will remain relatively unresponsive to the potential for harm that must be averted especially as larger and larger segments of the population are outfitted with in-the-body technology for medical or enhancement reasons. One day there will be a crisis—perhaps a wide-spread communication virus that disables RFIDs, stimulators or pumps, a hacker who changes heartbeats, a device that begins to act autonomously, so it cannot be controlled or shut down. Individuals critically dependent on the technology will become seriously ill or die, and those dependent on the technology for enhanced intellect or skill will be suddenly brought back to vintage 2011, a condition completely untenable to a society relying on the new capacities and resources. Suddenly, government regulators will scramble to enact new regulation on everything from how to deal with in-the-body availability, maintainability, and safety, to who shall pay for the enhancements, to laws that regulate matters of ownership and consumer protection, and how to shield the population from new scams and yet to be conceived criminal activity.

When the time comes that in-the-body technologies communicate with the outer world, issues will arise over access—who has it and for what purpose. Engineers will have to develop communication protocols and decide appropriate levels of encryption to prevent hacking, yet allow unencumbered access by the medical community. Regulators will impose licensing requirements on entities that control data collection and communication servers. Congress may have to establish agencies to police privacy and decide if medical device enhancements will only be provided through government licensed entities, such as hospitals or special techno-hospitals.

Agencies, such as the FDA, may eventually find that they are immersed in the regulation of software, controlling who will supply medical device software, software updates, and who must pay, or who decides when subscriptions to the software can run out. These matters invariably implicate insurance, health care, patents, copyrights, licensing, product safety and other consumer laws.

Once enhanced technology communicates outside the body, issues will emerge regarding viruses, bots and spam, and other evils and ills that befall the average computer user. Imagine the consequences of a deadly computer virus being let loose on a medical network, a virus that causes vital, life-preserving apparatuses to malfunction. What kind of policing system is needed, what are the costs of defaults, of violations in security? And for those that perpetrate crimes against in-the-body devices, what are the penalties?

The Medical Device Security Center is operated as a private partnership between Beth Israel Deaconess Medical Center, Harvard Medical School, the University of

Massachusetts Amherst, and the University of Washington. It has as its mission the balancing between security, privacy, safety, and effectiveness for next-generation medical healthcare devices. In one study researchers investigated whether hackers could gain wireless access to combination heart defibrillator/pacemakers that allow doctors to monitor and adjust operating parameters remotely. The experiment simulated the effects of a "hacker" using a readily available commercial programmer and a software radio that can be purchased on eBay. The test successfully demonstrated that a hacker could reprogram the device, shut it down, deplete its battery, or deliver jolts of electricity that could be potentially fatal to a would-be patient. The researchers said they had been able to obtain personal patient data (name and diagnosis) and medical telemetry data by snooping on signals from the embedded radio. In the future device designers will employ mechanisms referred to by the researchers as "zero-power defenses" that audibly warn patients of security breaches and will use cryptographic power authentication protocols to prevent unauthorized access. "The risks to patients now are very low, but I worry that they could increase in the future," said Tadayoshi Kohno, a lead researcher on the project.

In 2011 an independent security researcher spoke before Black Hat 2011 and revealed vulnerabilities with the implanted insulin pumps worn by diabetics, which would allow attackers to remotely control dosage rates. Currently Federal regulators have not made public any security breaches of the kind mentioned here; however, as in-the-body devices proliferate among the population, the security issues can only worsen. Unlike hacking into financial or medical records, where the loss can be catastrophic, but are ultimately economic or privacy crimes, these crimes are perpetrated on an individual and in the nature of an assault. Additionally, the ability to hack into in-the-body devices presents a life-threatening breach that might not be merely directed at a specific victim, but to an entire population, ordinary people and world leaders included.

The increasing numbers of medical devices that contain electronics will become increasingly complex, both from a device and a system standpoint. Evaluating devices and systems for efficacy and failure modes will require new specializations. Doctors trained in electrophysiology, bioengineers and medical technicians will need to have an in-depth knowledge of electronic medical devices, embedded systems, systems engineering and communications. Eventually the medical profession itself will be required to train its doctors in the skills that will allow them to install not only in-the-body technology, but to appreciate their limitations, upside and downside potentials, and to diagnose malfunctioning units. Doctors will reach the point where they must be as trained in this kind of technology as they once were in microscopes. The breadth of engineering disciplines required to accomplish such a mission poses a significant challenge to the medical establishment and particularly its educational institutions.

Systems comprised of medical devices that interoperate in the single-person environment can be divided into three areas: 1) sensors, stimulators, actuators, and pumps that deliver therapy and make measurements; 2) network architecture, including telecommunications that interconnect the devices both in-the-body and to the outer world; and 3) mechanisms to program applications using the interoperability architecture. These interoperable systems will allow for sharing of information throughout networks both wired and wireless. Therapeutic devices will be tailored to a range of specialized clinical situations, home care, hospitals, the battlefield, and portable applications. Standardization of protocols will be required even though most will operate through local area networks (LAN), wide area networks (WAN), and access the Internet, where diagnostics for hardware, software, and biological function will be routinely run on patients.

Companies yet to come into existence will focus on the challenges and benefits of interconnecting medical devices from different manufacturers, into a seamless medical device "plug-and-play" network. Products that have been historically vertically integrated into a system by a single manufacturer will become components in a larger multivendor system, where larger corporations, will offer bundled services. They will establish centers to accept calls from in-the-body technology recipients complaining that the device or system is not functioning according to specifications or other expectations. The call center will attempt to remedy the problems that are not deemed life threatening, before sending them up the chain to experts that have training in computers and medicine.

Teams will be formed of industry engineers, academic researchers, regulatory personnel and clinicians from around the world, who will be enlisted to study the clinical complexities of interoperable systems (the hazards and risks) before anything is market approved. For example they will help develop policies and regulation to prevent problems in unauthorized device access or the lack of effective virus protection (problems currently unaddressed).

Typically these devices maintain a bi-directional radio communication with a remote computer utilizing the Medical Implant Communication Service frequency band between 401 MHz and 406 MHz. The maximum power transmission is quite low at 25 microwatts, providing a range of a couple of meters to reduce the risk of interfering with other users on the same band. The maximum bandwidth is 300 kHz, which makes these devices low bit-rate systems, compared with WiFi or Bluetooth.

Key areas of concern for FDA regulators will continue to be: Electrical safety hazard issues in medical electronic devices; failsafe medical electronic device designs and reliable manufacturing; safety issues related to batteries; interference from errant electrical transmissions; the error free integrity of data acquisition; interoperability and compatibility among different vendors and subsystems, both in the body and in the outer communicating world; dependable and accurate modeling and simulation of electronic systems.[152] Each of these presents enormous challenges currently, but will multiply in complexity for a future shaped by in-the-body technologies coupled to telecommunications. Questions remain as to whether these kinds of systems will become part of a larger FDA regulatory oversight.

Ours is the first age in which many thousands of the best-trained individual minds have made it a full-time business to get inside the collective public mind.—Marshall McLuhan, *The Mechanical Bride: Folklore of Industrial Man.*

THE ANATOMY SOCIALIZED AND COMMERCIALIZED

Like the paradox that light exists as both particles and waves at the same time, we too live in a paradox—existing both as individuals and as part of a social collective. Physiologically, consciously, tribally, we each combine with others to form ripples on the surface of a wave,—unmoored and atemporal, vast and ongoing, the sum of us defining a future architecture, the wave of humanity, not the individual in and of itself, but the whole.

In some ways, like the influence the moon has on its tides, technology has always worked to pull together an ever-growing tide of the whole of humankind. We began modestly with tipped spears and rough hammers, chisels, hoes and plows, bridges and boats, ancient aqueducts, sewers, telegraphs, telephones, radios, computers, the World Wide Web, email, the iPhone, Facebook. Each milestone brought humankind closer and closer. We began with a steady stream of know-how that divided into uninterrupted tributaries of invention merging into a determined river, itself spawning new tributaries along the way.

We soon will be at a point in our technological history where we will consolidate the intimacies of our individual anatomies (e.g., whether we are suffering from the common cold, the flu, how often we have sex, defecate or lose our temper) into cloud storage that will then be shared with unforeseen interests, some representing the medical community, others representing device and drug manufacturers and government agencies, such as the Center for Disease Control. Eventually this collection of information will pour into an ocean of epidemiological statistics, churned by titanic supercomputers, stored and retrieved in physiological databases that will monitor society from 50,000 feet to insure that it remains safe from the communicable diseases that have haunted humankind from its inception. At the moment we fall into the fold of this vast autonomous physiological organism, the species will have been totally immersed in the river of technology that began as a trickle at the dawn of civilization.

Long before we reach the point of a collective database for all that ails society, in-the-body devices, some as simple as RFID implants, will have routinely communicated to the outer world, e.g., transferring vital medical data to physicians. But eventually this same technology will gather information about personal habits—diets, drug stores, local bars frequented—and communicate this, too, to a non-medical outer commercial world. A company may offer products based on the known activities of an individual from a record of participation in daily exercise, sports, travel, or dangerous activities (as determined by insurance-like institutions). As the data is interpreted by a vendor, it may suggest products to alleviate some symptom (potential or actual), or to enhance a particular performance, or avoid a possibly harmful consequence of an activity. Someone who travels frequently by cruise liner might be offered a discounted purchase of Dramamine or if they travel by air something for anxiety.

When computers process information, the signaling resembles the structure of language, ordered and meaningful in the intended context. If the data is a medical or financial record about us, it is interpreted according to institutional rules established by the relevant recipient: medical community, insurance company, employer, and bank. In-the-body technology computer-communication does not represent an ordinary record about us; it is at its foundation the functioning of our anatomies. It is a copy of the internal workings of our bodies. In the future, an essential part of us, the signals that keep us alive may reside somewhere in cyberspace, among a mass of data, in a community of disparate interests controlling remote databases. As the foregoing suggests, these will be owned and controlled by vast corporate complexes whose main concentration will be on competition, not the consumer.

The saddest aspect of life right now is that science gathers knowledge faster than society gathers wisdom.—Isaac Asimov, *Isaac Asimov's Book of Science and Nature Quotations.*

CONCENTRATIONS OF POWER

The potential for harm that is brought about by in-the-body computers connected to vast databases raises the question as to whether and who will regulate such technology—the medical community, government or a self-regulated industry. Clearly there are safety and privacy issues in the telecommunication of medically sensitive data. For example, just as cookies, viruses, and adware are to the personal computer, in-the-body computers can similarly be compromised, not to defeat their intended purpose, but a mentioned earlier, to provide third parties with private information that is extraneous to the primary function. This data is of some benefit to the user, and is also useful to the manufacturer, since armed with certain information companies can improve the product, make it more reliable and cheaper to produce. Some companies will develop deluxe models and others economy models. Yet others will look to squeeze out efficiencies, compromising performance or safety.

As previously pointed out current technology permits physicians to maintain and alter, over a computer link, the rates of drug deliveries or the rate of electrical current excitation that a device such as a pacemaker or brain implant delivers. As technology advances new standards of care in the medico-technological community will handoff autonomous control over these devices to remotely located servers. As we operate in a cyborg-form a constant stream of tweets carrying sometimes vital and sometimes non-essential queries will keep our platforms responding to the external world (much as our Internet browsers do now). Most of us have woken to the morning message that our computer has been recently updated. All of this will require that the central computers controlling these operations be regulated to ensure that their human subjects are safe and secure.

In addition to FDA regulation of in-the-body devices, will the central computers themselves become part of the FDA regulatory scheme? The FDA deals with combination products which include, among other things, products largely comprised of two or more regulated components, i.e., drug/device, biologic/device, drug/biologic, or drug/device/biologic, that are physically, chemically, or otherwise combined or mixed and produced as a single entity. On the FDA's official website it states that,

> Because combination products involve components that would normally be regulated under different types of regulatory authorities, and frequently by different FDA Centers, they raise challenging regulatory, policy, and review management challenges. Differences in regulatory pathways for each component can impact the regulatory processes for all aspects of product development and management, including preclinical testing, clinical investigation, marketing applications, manufacturing and quality control, adverse event reporting, promotion and advertising, and post-approval modifications.[153]

It is not clear whether the FDA is positioned to apply its expertise to the world of internal medical devices integrated into an external world of telecommunications.

As we carry these issues forward, we face uncharted territory where property rights and equitable rights may not have the same mosaic they had in the land of standard issue bodies, those without benefit of retrofitted parts. In the eighteen hundreds one celebrated case dealt with a landowner and the responsibility he had for maintaining a potentially dangerous body of water on his land. In that case a reservoir flooded a nearby mine.[154] The court held the owner responsible for the consequences of the act regardless of how carefully he had maintained the body of water. The mere possession of such a potential force was enough to hold him accountable for its release. In today's world, the mine of yesteryear has been replaced by the technology of today. What happens when an in-the-body prosthetic update broadcast to millions goes haywire or perhaps contains a virus? Those who own the technology must be held accountable for the consequences actions and failures to act. To what commercial, medical and legal standard will those who supply enhancements be held? Suppose the guaranteed gain in human intelligence falls short, or the advertised extended human lifetime is off by a decade or two, or the warrantee as to an invulnerability to a virus has been breached? As in-the-body technology becomes as common as body piercing, significant warranty issues will likely cause company managers, regulators and lawyers many sleepless nights.

Modes of consumer protection that took hundreds of years to develop, may no longer function equitably in a world where the human anatomy takes a sharp turn and needs novel forms of protection. In an age of in-the-body computer networks, property takes on a new significance because of the intangible features of information it now must accommodate. Examples of intangible property might be the databases used to feed the processors located somewhere in the body. Who will fund, own and thereby control the databases that communicate with the in-the-body processors? Who will control the network insofar as maintenance, updates and security? Under what circumstances might service to the individual or more pointedly the embedded sensors, be terminated or disconnected from the network? One might consider myriad possibilities along these lines. However, each of these examples has in common an alteration to our social reality and sense of ethics, which necessitates corresponding governmental policies and bioethical assessments.

How medical device products and processes are monopolized by patents and trade secrets is not something widely appreciated by the consumer, but obviously plays a significant part in who is afforded the opportunity to purchase and how much they must pay for the products. In fact, protecting aspects of product or process through the patent system is initially at least secret. Trade secret aspects of a product or process remain secret, indefinitely. Patents and proprietary secrets erect an impenetrable wall that allows practically unfettered economic power in the marketplace. Especially in technology law, where the property right at stake does not reach into the public psyche, much about the product's impact on society depends on the protection of its proprietary methods, and more specifically such things as the particular recipes where drugs are involved or algorithms where software is involved. Anticompetitive advantage is propagated and advocated by a relatively small minority of special interests: the particular industry, industry associations and the patent bar, which tend to help navigate the policy, legislative, and bureaucratic apparatus into preferred and lucrative directions.

Ownership in the intellectual property will affect availability, price, performance, safety, and choice. Without regulation, commercial interests might well run counter to bioethics, raising issues in autonomy and distributive justice, primarily when the lack of affordability restricts access. This effect is not much different from the experience some have when they discover they cannot afford medical treatment or drugs: deprivation may mean the difference between life and death.

Technology... is a queer thing. It brings you great gifts with one hand, and it stabs you in the back with the other.—C.P. Snow, *New York Times*, 15 March 1971

ADMISSION FEES

In many countries, even in life or death situations, one may not be entitled to medical treatment, unless they can afford to pay for the procedure. In 2010, it cost an average of $8,402 per person for health care. And as it presently stands, inflation produced by patent monopolies from medical drugs contributes a fair share to the overall cost. According to a recent study: "Health care experts point to the development and diffusion of medical technology as primary factors in explaining the persistent difference between health spending and overall economic growth, with some arguing that new medical technology may account for about one-half or more of real long-term spending growth."[155]

Should an economic model be imposed for in-the-body technology, different from the model used for most commodities? New in-the-body technology finds itself in an altered industrial, trade, and commercial category than do the technologies of steel, minerals, oil, and widgets. Yet in respect to drugs, in-the-body technologies, transplants, and even new medical procedures, we witness government essentially taking the same laissez faire attitude it takes in non-medical sectors. But in-the-body technology is not the same as most commodities, and its equitable distribution must not be lost. Economic models need to consider patenting, licensing, pricing and universal availability.[156]

We associate equitable distribution with the idea of ensuring that society allocates through political and economic means resources sufficient to enjoy life, liberty, and the pursuit of happiness. In the U.S. this ideal roughly equates to the American dream. When the available wealth and resources are unfairly concentrated in one or another segment of the population, prevailing legal and moral forces will tend to adjust the most severe inequities.[157] A vast store of precedence shows U.S.

jurisprudence can be prolific in setting things right between private interests (in this regard, the U.S. exceeds the world in civil litigation). Patents represent quasi–public interests, and historically, the matter of patents involving public interest versus private interest has been considerably more divisive and more stubborn in coming to a resolution (perhaps due to our libertarian and conservative provenance).[158] At the other end of the spectrum, Karl Marx believed that wealth unfairly accreted to the capitalists at the expense of the laborer. Adopting this as dogma, Communists conquered half the world in the name of equitable distribution. In consequence of their hapless experiment, at least as to the Soviet Union, the moral, social, political, and economic state of the world was made poorer, not richer. How and where do legislators draw the just boundary?

The failure of one economic system or the success of another does not provide ethical guidance to a society attempting to balance its values for life, happiness, and the technological potentiality for the fountain of youth. How does a society balance these beneficent against the maleficent of some futuristic chimera, runaway malignancy, or human species largely under the control of in-the-body technology?[159] Essentially, must we make some kind of Faustian bargain to advance science and technology and will this run counter to our values? Laurie Zoloth points out that in the Jewish ethical-legal tradition the "... framing questions are those of obligations, duty, and just relationships to the other, rather than protection of rights, privacy, or ownership of the autonomous self." She writes further that much of American bioethics is rights based and "...relies on a model of intricate semilegal contracts..." [160] Employment, non-compete and confidentiality agreements, patents and the appertaining assignment of rights and technology licenses are examples of how this plays out in commercializing biotechnology.

Patents especially set the stage for how businesses deal with their investments, competition, pricing and distribution. Fundamentally, patent law strikes a bargain between the patentee and the government, whereby the later grants a 20-year monopoly on an invention in exchange for the patentee disclosing the full particulars of the invention. Once the government grants the patent, no one can practice the invention without a license. The patents that issue around in-the-body processes have the effect of removing from the commons, in greater or lesser degrees, the underlying technology when it relates to a particular family of ideas. The effect of a strong patent position contributed substantially to rapid and enormous success of enterprises like as General Electric, ATT, IBM, Microsoft, and Apple. We might generally choose to ignore this phenomenon, except when its practice affects an organ responsible for the health and welfare of a person.

When exploring policy pertaining to emerging technology, numerous factors determine where it is heading, but the main ones are feasibility of the technology (proof of concept), reliability, and whether there is a threshold market to support a profit. In many cases, the route advances according to technical and commercial conditions, often without hearing one or another voice crucial to understanding all sides of the debate, especially those that may question the safety or the technology's potentially harmful effect on the consumer.

And now, though feeble and short-lived, Mankind has flaming fire and therefrom learns many crafts.— Hesiod—Edith Hamilton, *Mythology*

INVENTIONS ALTER SOCIETY UNPREDICTABLY

Who could have predicted that Eli Whitney's cotton gin, which served as a stark example of creating an economy that prospered for many, but greatly influenced the acceptance of slavery? Following commercialization of the invention, circa 1800, plantations needed larger work forces and expanded slavery in both numbers and territorial reach. With each succeeding decade cotton production doubled. Increased crop production motivated other inventors to create new machines to spin, weave, and fabricate new cotton-based products. Transportation methods, both for the raw materials and the finished goods began to change. Significantly, the cotton gin increased the demand for low-cost workers, which caused an increase in slave importation (in 1790, six slave states existed; and by 1860, fifteen existed). From 1790 until 1808, when Congress banned the importation of slaves, Americans condemned over 80,000 African slaves to labor on ever-larger plantations. And ultimately, the enslavement of tens of thousands led to the Civil War, in which hundreds of thousands were killed by the latest war-technology.

Of course, not all inventions lead to social upheaval, suffering, and war—many reduce suffering or advance civilization in positive ways. Yet, the greater inventions of the modern era often have had side-effects that were not wholly intended or for the greater good. Historians can pinpoint specific technological advances to explain how social reality changes, i.e., what things appeared to mean, before and after. But we are not here to report on history, but on the present and what it means for the future. Today, every year surpasses the last, and the time between basic discovery and practical application continues to compress. How can we better estimate the effects on our society? This question begins and ends on the imperative to better understand what things mean as our society undergoes a storm of technological revolution.

The definition of technology itself has mushroomed encompassing not only things of substance, but also the very ideas about that substance. Just as our forebears had to recalibrate their cultural compasses when Galileo announced that the world was round, we now need to better understand the expanded notion of technology and perhaps recalibrate our psychological, social, and moral views as well. For the human anatomy stands at the forefront of an era when it will host an array of embedded computers, each having the power of a network of 1960s computer—the kind that once occupied several thousand square feet in air-conditioned rooms.

The past twenty years have witnessed the new sciences of genetic engineering, nanotechnology, cognitive neuroscience, practical applications of artificial intelligence, and a cornucopia of curative drugs and biomedical devices. Although the broader outlines of what was coming threaded through our lives from the '70s through the '90s, no one could have anticipated the specific products of the 2000s, omnipresent WiFi or GPS, personal data assistants/mobile phones (iPads, iPhones, Androids), embedded RFID chips, 3-D TVs. No one predicted the estimated 8.6 million robots (and are doubling in numbers every three years), Next Generation Sequencing to profile specific cancers genomes for individual treatment response, or pharmaceuticals based on DNA science.[161] New cultural patterns established by the likes of Google, Facebook, Twitter, and Youtube, will morph over time and eventually be replaced by patterns borne of technologies yet conceived.

Aside from its cultural impact, the steep curve of technological innovation produces an unprecedented world wealth that manifests in the form of intellectual property, creating, as mentioned earlier, an effective market monopoly on products that are essential to staving off disease or non-essential but nonetheless enhance life's experiences.

Current explosive technological growth, which includes the pharmaceutical and medical device industries, distinguishes itself from the past by the highly competitive immediate monopolization of ideas behind each new technology. For example, the number of patent applications filed in the U.S. has jumped from 177,830 applications filed in 1991 to 535,188 in 2011. This is only explained by the facts that more than any time in history, patents epitomize the currency of competition, and thus significant economic control over the distribution of products.

Today's innovations grounded in patents come with strings attached to licenses, cornered markets and prices that frequently generate revenues far in excess of what is fair and equitable to return investments. In this modern era we seem to blithely accept merchandise and services with all the attached strings, with little thought of how it all combines to change the culture, first one decade to the next, then one year to the next, increasingly compressing time, washing away the last generation of products, replacing what we do at this very moment—engaged in activities our grandparents would not recognize. Most of us hardly notice what is happening before our very eyes.

Consumption today is significantly driven by the intangible processes that run the Internet, computers, telecommunications, monetary instruments, (derivatives, futures, and monetary policies) and increasingly medical care. These developments, largely created by mathematicians, scientists and technologists are seized upon by corporations that then construct novel ways of doing business such as delivering medical care, each spawning new market opportunities, new competitors, and political adversaries.

To what extent do novel ways of enhancing the anatomy analogously alter the traditional ways of delivering medicine? A medical community with ready access to the body through communication mediums will change our once direct patient-doctor relationship into one separated by a wall of technology, where the technology will serve not only the patient, but the interests of those who participate in its commercial success, its maintenance or those that create abstract social-medical systems that might not completely serve individuals. Certainly we have seen how the explosion in revolutionary pharmaceuticals changed economics and delivery models for drugs, e.g., the

Medicare Prescription Drug Act of 2003. In the U.S., as opposed to other places in the world, drug companies now routinely advertise directly to the consumer. What happens when the 30 or 60 day drug prescription is extended by months because a small drug inventory resides in a miniature warehouse stored inside the body, resident in an anatomical cavity, and delivered to the precise location upon receiving a signal from a drug management system? Will these systems be connected to your doctor for constantly updating your medical record or local pharmacy for drug refilling, billing, and record keeping? If you were to desire a new drug, your doctor might only need to interrogate your anatomically embedded electronic computer from his or her office and then key into his or her computer a prescription to the pharmacy. It is likely that a widespread use of in-the-body computers and communication devices will allow examinations from standard computer terminals. Doctors may look at the data collected during so-called medical examinations hours or days after the fact, unless alerted by a programmed semi-autonomous "intelligent agent" that they look sooner. In some cases a computer will analyze the data well before a doctor sets eyes on the raw data. In some ways this is not too different from the local medical laboratory analyzing blood and reporting the results to the physician, but on a different scale. And, in time, as the computers get smarter, there may be a tendency to relegate more and more decision making to the computer's "intelligent agent", removing in some respects the physician from routine doctor's orders or prescriptions. These are probable scenarios, but because of the rapidity with which technology is moving, policy makers need to better understand the potential for mischief and put them on today's agenda to insure that the integrity of the medical system does not devolve into one which as a practical matter involves only the patient and an "intelligent agent."

It has become appallingly obvious that our technology has exceeded our humanity.—Albert Einstein

WHO WILL OWN IN-THE-BODY TECHNOLOGY?

Prosthetics such as a dental bridge, pacemaker, or those that replace a limb or an organ embody two kinds of property interests. First is tangible personal property, which grants the right to do with the article what the owner desires. He or she could sell the item as freely as selling a table or chair. The exception is that body organs (e.g., heart, kidneys, liver, lungs, pancreas, intestine, and thymus) and tissue transplants (e.g., bones, tendons, cornea, skin, heart valves, and veins), whether in a body or awaiting transplantation, are not allowed to be sold.[162] A second kind of property, relevant to this discussion, is intellectual property, an intangible property right, which applies to the trade secret, invention or tangible expression (such as software) that underlies a product or process. The owner may assign ownership of the intellectual property or they may sell the tangible product or both, because they are, under law, deemed separate things. Another possibility is that rather than assigning or selling ownership, they may license the intellectual property for some purpose. The most common example is someone who purchases a book from an author or publisher. The purchaser owns the book, can read the story a hundred times, and sell the book when they are finished. But the story remains the property of the author or the publisher. Consequently, the owner of the book cannot make a copy of the book, because they do not have rights for copying the story. Similarly, a producer may acquire a license to the story for the limited purposes of making a movie, but otherwise does not own the story. Title to the intellectual property, in the case of the writing, does not pass to the purchaser of the book. Unlike outright ownership, intellectual property licenses generally restrict a licensee's options regarding changes or updates, unless approved by the owner. What is said about the book and the story applies equally to computers and software, and hence to the medical devices that constitute many of today's more sophisticated prostheses. Title to the intellectual property, embodied in a prosthetic, does not generally pass to the consumer.

The scheme of dividing the two kinds of property—tangible personal and intangible intellectual property—works well for most things, such as widgets, movies, books, and software, but it may not work well for articles that we become critically dependent on, such as an in-the-body medical or enhancement device. If someone has an in-the-body prosthetic and it needs updating, the license holder can restrict who provides the update or deny the update, and may in many cases, charge the consumer what it desires for the original article or the future updates. As Lyotard wrote "... Knowledge is and will be produced in order to be sold, it is and will be consumed in order to be valorised in a new production: in both cases, the goal is exchange. Knowledge ceases to be an end in itself, it loses its 'use-value.'" However, some in-the-body technology, such as the kind, without which the recipient would die, constitutes another form of "life" (as for instance a mechanical heart might be considered). Should it not be the absolute property of the recipient, updated as needed, no strings attached? As indicated, this is not generally the case, since almost every new device or biotechnical process is bound by a patent or copyright and is subject to a license, each with restrictions as to the legal right to make changes, use alternate supplies for updates or prohibit reverse engineering.

That being said, a radical change in current law, such as a prohibition or moratorium on patenting or licensing in-the-body technology potentially would have ruinous consequences for scientific research and medical innovation. Scientific research, whether private industry or university-based, is increasingly funded by investors looking for a return on their investment. The current and rather colossal research expenditure that supports medical device and biotechnology inventions will not occur without the assurance of commercial incentives to pursue research and development of new diagnostics and biologics. Yet, there is something fundamentally amiss with treating in-the-body technology like all other technology, especially when the technology is an essential adjunct installed inside the body to keep it alive.

Changing attitudes toward commercialization and ownership of basic scientific research and the analogous protection of all manner of invention or expression through the patent or copyright system, calls for studied consideration as to where we set boundaries and if a new ownership paradigm might better serve the interests of patients and competition. One might argue that in today's legal-commercial world, intellectual property ownership is unhinged from core human values. Courts and policy makers see the matter as one of economics. If more broadly treated, as I suggest it should be, intellectual property policy ought to consider not only the narrow goal of rewarding inventors and the commercial interests that invest in these technologies, but also the underlying purposeful nature of the products that keep people alive. This is not novel. Some inventions are considered "injurious to the morals, health or good order of society" and are therefore not patentable. [163] Government in these instances treats certain kinds of invention differently from others. In the case of inventions that are life-sustaining, government should take a fresh look at what ownership means and consider reforming the intellectual property laws to serve not only the commercial interests but the interests of the patient-consumer.

...as soon as we take the final step of reducing our own species to the level of mere Nature, the whole process is stultified, for this time the being who stood to gain and the being who has been sacrificed are one and the same.—C.S. Lewis

RATIONING THERAPEUTICS

The year is 2060 and Eve's grandmother Sarah has just turned 125. For her birthday she has decided that it's time for a cyber-lift, the term used by cyber-doctors for those who want a mid-life make over. The procedure begins with an examination to assess her state of health and, if she is relatively healthy, to inventory her in-the-body technology. Rebekah, her physician, has decided that she would benefit from a new computer set—the array that keeps a half-dozen major organs running like a European sports car, and which include the main server that collects statistical data and communicates with the outer world. Her cyber-functionality concerns nearly every aspect of her life. Internal processors keep organs that have reached their evolutionary life cycle functioning through stimulating electrical currents—some coded for use with the brain's circuitry and others to supply the regularity of a time-piece to parts of the anatomy that need to keep a regular beat. A system of pumps keeps her lungs from atrophying and blood glucose from spiking.

Aside from the apparatuses that keep Sarah alive, she also has installed a system of processors that provides enhancements to parts of the anatomy to keep her sharp as a tack.[164] For example, twenty years ago she had a Megacorti six-core 256 byte parallel processor installed that links the frontal lobes (which are involved in planning, organization and other highly developed human abilities) with the parietal region farther back in the brain, and integrating information from the eyes, ears, and other senses—thereby increasing intelligence.[165] Rebekah informs her that the one she has installed is very old technology and needs to have not only a new processor but a new operating system and new applications that require a new central processor.

The third major system affected by the cyber-lift are neither life sustaining nor enhancement but economic.

She uses smart RFID technology that has been around for half-century and all retailers prefer the technology, so she has found it harder and harder to shop for goods, pay bills, and receive her dividend checks. The newer systems are fully integrated with the latest banking protocols and have an added feature that allows her to instantly access any utility in her home by just thinking about it. For example, she can adjust the air conditioning in any room, and turn on and off the television, lights, and security system, without lifting a finger.

Her doctor goes over several plans—Platinum, Gold and Silver—each offered by Pegasus, the leading provider of the cyber-lift system. Fortunately, the pumps and several of the off-the-shelf stimulators, like the pacemaker and the deep probe brain simulator, need not be changed out. Unfortunately, she does need a new central processor and a peripheral processor that is used mainly as an intelligence enhancer. She finds the Platinum and Gold plans too expensive for her budget. She inquires if there are other vendors that might perhaps offer similar performance at a lesser price. Rebekah tells her that there are only two companies in the world that supply the operating systems and only one that supports the communications protocols that she will require. They commiserate that the prices are kept artificially high, but that two companies have a lock on the patents for the latest technology. Additionally, like the telephone companies of the late twentieth century, they have cornered the market on licensing, both communication frequencies and operating channels.

Although she has been grandfathered insofar as the technology she currently has installed, there are rumors that the processors Sarah currently has will only be maintained for another seven or eight years. After that she is at risk, medically and commercially, unless Congress passes legislation forcing the two suppliers to extend the service contracts.

Dismayed by all this economic bad news, she asks just how serviceable the Silver plan is. Rebekah explains that the difference between the Platinum, Gold, and Silver has to do with the differences between the professional grade, the home package, and the starter package. Each of them is probably more advanced than what she is now using. In fact she would probably not need the professional model because she no longer works and it is mainly for those that need greater search capacities, faster response times and she says candidly, a greater intellect to compete for employment opportunities.

After mulling her options, Sarah decides that she can live happily with the Silver plan, and with that she schedules the day that the system's change-over will occur.

As the twenty-first century races ahead, discovery and invention patiently await the harvesting of scientific seeds planted in the preceding five hundred years. When fertilized by the economic largesse of massive corporations, in-the-body technology will produce a veritable cornucopia of products. All things being equal, this would portend good things for society. Nonetheless, nothing is free; everything comes with a price.[166] We must consider and make economic choices on subjects that lie in deep and wide moral furrows—how access to in-the-body technology might divide the species between, not merely "good" or "poor" health, but of better, worse or no enhancements, that will determine employment and commercial opportunities, increase intelligence, and elongate lifetimes for a those that can afford it. In some regard, choice will be limited rather than opened, mainly because only a few corporations will have the wherewithal to support the vast system. This is not different from the monopolization that AT&T has exerted over the telecommunications business for over a hundred years, or IBM's monopolization of the architecture of the personal computer, or Microsoft's monopolization of the operating system and applications programs. In the last instance, one other competitor has been viable with respect to personal computing: Apple.

When I graduated from college, Nobel Laureate Dennis Gabor—the man who developed the theory of holography— was the commencement speaker. I looked forward to introducing myself, as I had been researching artificial holograms using frequency multiplexed video and I thought I might engage him in some professional small talk. But ironically it was not Gabor or our tête-à-tête that left a lasting impression; it was a form letter from then President Richard Nixon that was stuffed behind the degree. He referred to a Greek parable, where a student tried to trick his wise master.

The student planned to conceal a bird in his hands. He planned to ask the old man to guess what he was holding and, if he guessed a bird, the boy would ask whether it was dead or alive. Should the old man guess dead, the boy would let the bird fly away. But, if the wise man guessed the bird was alive, the boy would crush out its life and open his hands to reveal a dead bird. The boy finally asked, "Is the bird alive or dead?" And the old man replied, "My son, the answer to that question is in your hands."

For people that require in-the-body technology to remedy a health issue, the subject of its availability becomes whether or not access is denied. While lawmakers debate such things as health insurance, millions of people remain uninsured in the U.S., and throughout the world others are deprived of access to lifesaving drugs because of the cost or other economic reasons,—such as patent monopolies. The issue of unavailability not only affects tangible items, such as in-the-body technology, but medical procedures. Access to medicine is a major economic debacle of our time.

Not too many years ago, Africa was the focal point of debates between the pharmaceutical industry and the government about the right to make drugs needed to combat AIDS. Pharmaceutical giant Glaxo Wellcome tried to block access to the less-costly generic versions of its top-selling AIDS medicine. The company claimed that sales of generic versions of its drug, Combivir, in Ghana would be illegal because they would violate its patents. As a result, the competitor stopped selling the low-cost version. [167]

After a fierce debate between the South African government and a few dozen pharmaceutical companies, behind which the U.S. government was steadfast in its support threatening trade sanctions, Merck, an American company agreed to provide AIDS medicines at cost to the developing world. [168] Query: What constitutes the developing world? Would the offer include Brazil, India and China? Would the offer include the Pine Ridge Indian

Reservation in South Dakota? For a company at least expecting some economic return it would be writing off the developing world's AIDS market.[169] Would this set a precedent for treating other diseases such as malaria or cancer or countless other pandemics? Not likely.

Following the September 11, 2001 terrorist attacks, the fight over lifesaving drugs was brought closer to home. During the anthrax attacks that followed 9/11, the public added the antibiotic Ciprofloxacin Hydrochloride or Cipro to its medical lexicon. As epidemic proportions of poisoning incidents unfolded questions were raised among public health officials about the sufficiency of worldwide and U.S. stockpiles of the drug. Estimates were that the government had a supply of Cipro for two million people—10 million doses according to the Secretary of Health and Human Services, Tommy Thompson. Congress quickly convened hearings and listened to government officials and industry executives comfort an anxious nation that production of the drug was being stepped up. However, from all estimates, it seemed that the pipeline would only be filled it the manufacturer worked 24 hours a day for 20 months. In the course of testimony, we learned that the German company Bayer AG held the patent for Cipro, and that it would only allow a select group of companies to manufacture to the new demands.[170]

While we learned that antibiotics, such as doxycycline and penicillin could arrest forms of the disease, Cipro was especially effective for treating anthrax that had been inhaled. Nevertheless, the Secretary stated that the government did not intend to break Bayer's patent and that it would be illegal to do so.[171]Although, others pointed out that there were legal precedents to procure whatever the government needs in times of crisis, the administration made a political decision, not to pursue this avenue toward the solution.[172]

Translated, the South African AIDs controversy and the Cipro case are about whether intellectual property rights trump the right to save lives through drugs, medical procedures, and someday, through in-the-body technology. In certain instances, governments operating under international agreements and treaties, such as the World

Trade Organization's Trade-Related Aspects of Intellectual Property Rights Agreement ("TRIPS Agreement"), and the Doha Declaration on TRIPS and Public Health permit the issuance of compulsory licenses in order to enhance social and public welfare, while protecting the interests of the owner of the patent.[173] However, if every innovation or discovery were immediately dedicated to the general welfare, we would be unlikely to experience the groundbreaking advances we see in technology. It has long been recognized that inventors and the entrepreneurs, who underwrite and commercialize invention, deserve to enjoy the fruits of their labors and capital. As such, legislators and those charged with the administration of our intellectual property laws must constantly balance the interests of commerce, scientific advancement, and the rights of all people to the produce of technology —,especially those advancements that affect our health and well-being.

Whenever we invoke the word "property", we invoke the concept of dominion. Qualifying the word "property" with the modifier "intellectual" does not alter that fact. As with all property, there come certain ideas about entitlement, benefits, and disposition. If I own it, I have the right to do with it as I please. I may rationalize my behavior in the name of freedom, benevolence or countless other value-laden shibboleths, but in the end if it is mine I can deal with the property as suits my pleasure. This holds true whether the "me" is an individual, a corporation, or the government. Therefore, if I truly own anything, I need not share it, nor do I need justify for how much I sell it, nor whether I maintain its integrity (e.g., I am not legally bound to supply updates and may prevent others from doing so).

Case in point: Dr. Samuel Pallin was granted a patent on a method for performing an incision in the eye during cataract surgery. The advantage claimed in the patent was that it reduced the requirement for suturing; thus, the healing process proceeded with greater effectiveness. Physicians, after learning about the innovation, wanted to use it but the patent limited that option. One physician used it despite the patent protection and Pallin sued for patent infringement. Ultimately Pallin lost the case, but losing is not the situation in the majority of instances.[174]

In August, 2012, the Court of Appeals for Federal Circuit ruled in Association for Molecular Pathology, et al v. USPTO (AMP case) that DNA sequences extracted from the body are patentable. BRAC Analysis helps doctors assess whether women are at heightened risk for hereditary breast and ovarian cancer. The patent income contributes substantially to Myriad's revenues, which in 2011 was approximately $400 million. Myriad does not allow competition, so that women must pay whatever Myriad charges and cannot seek-out second opinions. Although the case involved testing for breast cancer, the legal issue devolved into what was meant by "information" and whether it was protected by the underlying patent. When ruling in Myriad's favor, the court also took into consideration the potential impact that invalidating gene patenting would have on the life sciences industry. This case illustrates that courts are swayed by economic issues, and in some instances skirt the hard definitional issues of what is meant by such things as "information."

The Pallin and AMP case stands in stark contrast to the actions of Rosalyn Sussman Yalow winner of the Nobel Prize in Medicine and inventor of radioimmunoassay of human blood and tissue. In 1977, she and her team invented and developed an important method of chemical analysis, one still used in diagnosis today. Yalow and her colleague exhibited a rare generosity when they dedicated the invention to the public. Few companies that hold such rights—often worth millions—would ever consider dedicating so valuable a right to the public. Her dedication of the "use" was a moral statement that moved the matter beyond expressions of the market.

Every time I see an adult on a bicycle, I no longer despair for the future of the human race.—H.G. Wells

PROCESSES AND ARTIFACTS

Commercial and medical establishments of old were made from bricks and mortar, and once inside men and women serviced customer needs with a smile. Today, the portal is frequently not through swinging doors but through an Internet portal. Needs are serviced electronically through a seamless collection of computers and accessories that transport electrons according to certain logical rules that carry tokens of human expression —ones and zeros that transform into the words we see in our emails or webpages. In other words, information, literary expression, art, economic trading instruments, and technology now flow into a common artery to create new kinds of technological and social objects in everyday life, which during the short time the Internet has become widely available, has resulted in tens of thousands of lawsuits over a whole new class of rights, duties and obligations.[175] Once we recreate the Internet into a fully operational in-the-body network controversy and litigation will not abate, on the contrary, it will lead to massive disputes over yet another new class of rights, duties and obligations having profound implications for patients and society.

Inventions do not predict the future, but they infer what's likely to follow. Thirty-odd years ago, the Supreme Court ruled as permissible the patenting of what were essentially mathematical rules, referred to as software algorithms, provided they were part of an invention that produced a transformation, and a concrete, tangible result. Intended or not the effects of allowing the monopolization of these processes changed society for all time. Algorithm-centric patents led to the eventual acceptance of the business method patent, which protects how one might conduct a business (commercial, medical, charitable), over the Internet. Never in history, has there been such an uptick in inventions for forms of doing business. Tens of thousands of instances of a new class of social artifacts came into being with such household names as Priceline, eBay, Amazon, and Google. It turned the record industry and book publishing industry on its head, and it allowed the average citizen to trade in stocks, deposit and withdraw money from banks, search for doctors and accountants, or find their way from home to a strange location. Perhaps it most changed the manner in which we befriended people— through Facebook, MySpace, and Yahoo. These ways of conducting business did not exist prior to 1995—less than a generation ago. What is presently occurring is that there is a confluence of the software algorithm-centric patent and the bioengineered patent, combined which will again mint new social constructs. The next paradigm change will arrive when these social constructs, some as a consequence of computer technology, synthetic biology or a mix of the two are coupled directly into the human anatomy.

As engineers were designing the next generation robotic devices in 2010, we learned scientists at the J. Craig Venter Institute announced the world's first self-replicating synthetic genome in a bacterial cell of a different species. President Barack Obama requested that the President's Commission for the Study of Bioethical Issues immediately identify ethical boundaries surrounding this development. The Commission convened a series of public meetings in Washington, D.C., Philadelphia, and Atlanta, to hear and assess claims about the science, ethics, and public policy relating to self-replicating synthetic biology. In December 2010 the Commission issued its report entitled *New Directions-The Ethics of Synthetic Biology and Emerging Technologies*, which discussed in detail the ethical issues and the science as well as the potential benefits to society.[176] Can governments keep up with the change?

During the past 25 years, word processors, cell phones and the Internet have revolutionized the distribution of our intellectual output. These developments have fused the Information Age to the Age of Technology in a way that blurs the line between expressions (what we say) and technology (how we do it). If we treat technology as property, then at the interface, we must treat expression as property. What are the implications for a world where the technology qua information is part and parcel of our anatomy? Who might lay claim to this so-called intellectual property? Technologies that once sprouted from the physical elements of a material world are gradually replaced by technology steeped in the rules of social constructs. Property interests are no longer obvious; there are no wire fences. Today, we merely postulate some abstract idea and then stake our claim to its ownership. Such has been the concern over patenting DNA sequences, and software behind business methods. [177]

Inventions serve ulterior ends—some of which may implicate established values, such as the human form. I find it impossible to see any extant ethical issues in the invention of the doorstop. However, weapons of mass destruction, devices that invade privacy, dams, roads, automobiles and thousands of things are associated with our value system—e.g., the ends served. Still other inventions are ethics provoking, not solely because of the ends they serve, but additionally because of something discordant between its formation—e.g., the invention of a new *form of life*.

The patenting of new species—either human or animal —plays equally into a particular concern for the integrity of *forms of life*. The patenting of so-called disease genes are no exception. These patents generally claim a gene sequence (its code) associated with disease or risk of disease. In addition to covering all uses of the chemical sequences, these patents also claim methods of diagnosis by identifying a specific patient having the disclosed genetic alleles, mutations, or polymorphisms. Although there do exist institutions that freely disseminate such information, many gene sequences (and proteomes to follow) are being kept as trade secret or are monopolized through the patent process. This, of course, limits society's potential to save lives.

Historically, codes occurring in science and mathematics (biology or computer science) were not protected. A new and useful structure created with knowledge of such codes may be. Expressed sequence tags (ESTs) are gene fragments that help scientists mark precise locations of such things as traits or diseases. [178] But because they play so fundamental role in the biology of genetics, there has been repeated opposition at home and abroad that EST should be consigned to the category of research tools. The analogy is equivalent to allowing patents for the basic logical circuits that a computer depends upon to run its programs.[179] If the analogy has merit, the extension to the analogy to logic circuits and their mathematical representation, the basic Boolean equation, would mean that to patent an expressed sequence tag, takes the analogous biological equation out of circulation for the term of the patent.

Today, one can patent genes, gene therapy, transgenic animals, expressed sequence tags, antisense oligonucleotides and single nucleotide polymorphisms. Patenting gene fragments has reached numbers in the hundreds of thousands.

In the 1970's, scientists were beginning to discuss the domestication of microorganisms. In *Future Shock*, Toffler referred to how this would alter our food supply. As related to genes, we have gone well beyond both these points in domesticating the human gene. This means that we have learned to train it, to make it work for us, and to transmute its properties to serve purposes far beyond our wildest predictions. We have experimented on animals this way for thousands of years. The wolf of yesteryear looks nothing like the chihuahua next door—yet surprisingly they are essentially the same animal—except as to their look, physical capacity, and behavior. Someday, we will construct variations of the human design that may not look like us, have different physical and mental capacities, and behave differently. They will be fabricated to a specification to specialize their process. [180]

Through its rulemaking powers government can create the means through which the community can expresses its collective preferences. It can assign function and purpose for a particular class of inventions: such as new life forms. Unless we converge into a broad consensus on these issues, private institutions will form our social reality—at least on the issue of owning life forms, perhaps life itself, when the life itself becomes powered by technology.

As we look toward technologies that will not only save lives but those that will enhance intelligence and skill, we will need laws that consider the ethical prescriptions that are often lacking in laws and that in many instances disproportionately reflect commercial interests. For example, we must answer the question of whether to permit patenting all manners of development or whether there are other ways to compensate inventors or investors.

Science helps us before all things in this, that it somewhat lightens the feeling of wonder with which Nature fills us; then, however, as life becomes more and more complex, it creates new facilities for the avoidance of what would do us harm and the promotion of what will do us good.—Johann Wolfgang Von Goethe, *The Maxims and Reflections of Goethe*

THROUGH THE LENS OF A CYBORG

The year is 2075 and Adam, a twenty-five year-old living in Seattle, recently married Eve, who looks no older than Adam's twenty-one year old sister, but who actually just turned 60. Adam and Eve are considered transhumans according to the census bureau, which keeps track of such things. Not everyone is transhuman, partly because it is more expensive to outfit a child with the latest technology, than it is to send him or her to college, but Adam and Eve's parents decided, when they were born, to provide the latest complement available of in-the-body enhancers in the hope that their children would have all the best advantages. But, over the course of several decades, the technology and its platforms have improved dramatically, and it is nearly impossible to keep up.

Adam and Eve learned last week that they will be the proud parents of a baby boy by the end of the year, so the couple has begun to select the language accelerator technology that they will have installed before the baby comes home from the hospital. They will add additional components as time goes by, but they have decided that the ability to speak and read early is vital, if their son is to have a decent start in the increasingly competitive world.

Adam works as a bio-programmer with Nanocomp, one of the leading bio-computer manufacturers on the West Coast. A few weeks earlier he had been given unsettling news that the company was being looked at as an acquisition candidate by Cognosoft; and that the new employer may only keep employees that have a particular installed technology. Adam discovers his older embeds are incompatible.

The Mark 7 computer at Nanosoft connects to Adam's central imbedded server each month to provide updates to his technology and to capture his state of health—the first ensuring he has the latest tools for the job and the second to keep him healthy, while reducing insurance premiums and medical expenses. This past week, the Mark 7 generated the usual report detailing Adam's inventory of in-the-body hardware and software, but additionally confidential information about Adam, which he authorized in advance, when Cognosoft indicated it was needed in connection with the potential purchaser's due diligence.

While reviewing Adam's internal systems, the prospective employer learned that he contained a decade old version 4.6 cerebral processor from IB-X that contained a suite of memory storage; a version 5.8 chip set from Bioworks, Ltd., that regulated his sleep and imported physiological data during the evening hours, and a version 10.7 metabolic processor from Genefutures, LLC that kept track of each of the twenty major organs and communicated these weekly in a cryptographic code to this physician; and finally a version 1.2 commercial engine, supplied by Veritas-AG, Inc. that retained a trusted identification codes, as well as the key code to his internal "safety deposit box"—a nano-storage device that maintains everything from his bank account to his retirement account.

A few weeks pass, before Adam learns that his central processor is incompatible with Cognosoft's systems. What this means is that the adjunct memory bank, (a Google-like search cache), will not accept Cognosoft's protocols and that a changeover is not feasible, since the company can more cheaply hire the programmers that have systems compatible with theirs.

When Adam hears the news, he contacts IB-X and asks if he can change out his processor, and he is told that an entirely new implant will cost upward of $50,000. Adam recoils at the cost and decides that keeping his present job will not justify the changeover. In any case he and Eve were saving up, so that their baby boy would have the latest IB-X version language accelerator.

The World Transhumanist Association (WTA), founded in 1998, states it mission as developing an academically respectable form of transhumanism, "freed from the 'cultishness' which, at least in the eyes of some critics, had afflicted some of its earlier convocations."[181] WTA's founding document *Transhumanist* defines what post-human means:

> Posthumans could be completely synthetic artificial intelligences, or they could be enhanced uploads... or they could be the result of making many smaller but cumulatively profound augmentations to a biological human. The latter alternative would probably require either the redesign of the human organism using advanced nanotechnology or its radical enhancement using some combination of technologies such as genetic engineering, psychopharmacology, anti-aging therapies, neural interfaces, advanced information management tools, memory enhancing drugs, wearable computers, and cognitive techniques.[182]

We are born to share the benefits and liabilities of nature and nurture, and it is our parents who introduce us into the culture, exposing us to beliefs, values, needs, and attitudes. As we mature, they seek to give us the advantages of a decent education so we can live better lives than the former generation. But in the posthuman generation, the culture will be further determined by an internalized technology, which changes the cultural equation. Neil Postman, on the subject of culture in the age of modern technology, wrote: "... we are surrounded by the wondrous effects of machines and are encouraged to ignore the ideas embedded in them. Which means we become blind to the ideological meaning of our technologies."[183]

And we may be further blinded by the numerous versions of post-humans in our midst: versions 1.0, 2.0, some with replaced platforms, others with structural and functional revisions, where life at each update is not quite the same, or may in fact signify differences analogous to the kinds of differences that have plagued humanity since recognition that it inhabited Earth with others of lesser means, greater physical or mental prowess, different races, ethnicities and religions that offered strengths and weaknesses against the invader.

Again: Is there an invariable feature, an identity perhaps, that marks the *Homo sapiens* as constituted today, that will persist into the posthuman era? We have discussed technologies that are soon to be installed as relatively permanent non-organic and organic, internal adjuncts to our anatomy, furthering a mimetic transition, i.e., the widespread cultural or psychosocial acceptance of in-the-body technology. As indicated in the Introduction, this is the first stage, leading to a temetic end-state, i.e., where technology rapidly improves and replicates as it advances to stay ahead of competing technologies outstripping nature's slow ponderous process to adapt the *Homo sapiens* to an increasingly changing techno-environment.

In what way will we remain that life form that tracks its origins back 2.5 million years? Will we remain autonomous after being injected by RFIDs, infused with semiconductor sensors, modified by synthetic-DNA, and adapted with molecular and nano computers affecting our metabolic process, our mind, its perception, intuition, evaluation, and feelings? Will we retain autonomy in the sense of free will? Will we retain our identity, or will algorithmic-centric rules, built into new anatomical computers change or burden us new ways, in ways perhaps only remotely analogous to present limitations imposed by our physiology, culture, and psychological underpinnings?

And, if humankind is freed of its identity and autonomy, how will it know when it had reached the abyss, when it took that final transitioning step from *Homo sapiens* to posthumans? Is there anything about the sense of self that will warn it beforehand? Or, will our successors be informed afterward, in the historical annals of some future society that it had been engineered and manufactured—existing, in part, not wholly by the grace of nature, but that it had coevolved and adapted into a symbiotic relationship with a technologically based, artificially stimulated and coded, anatomical parasite?[184]

The best scientist is open to experience and begins with romance -- the idea that anything is possible.—Ray Bradbury, *Los Angeles Times*, Aug. 9, 1976.

QUANTUM SOCIETY

The year is 2084 and Adam and Eve have been married ten years. Adam eventually upgraded his internal systems and life had been good, but the extras they want are increasingly unaffordable, since any promotion Adam seeks will depend on the latest technology both in- and out-of-body. However, Eve decides on resuming her career, the one previously placed on hold when Scott was born. Like her husband and most of their friends, she uses an embedded cognitive processor, a chip the size of a black-eyed pea, which fits snuggly into the back of the neck just below her cranium. A titanium wire snakes into her cortex and carries thought-driven coded stimulations. But unlike the chips her husband or friends use, hers couples to her in-the-body main server, which communicates with the outer world to instantaneously retrieve the latest commodity trades. But in 2084, her version is ten years behind the latest version and this may affect whether she can effectively compete in the position she recently applied for with the WCT Corporation.

Things that we invent, our technology, tools, and appliances, and items that we consume all perform functions. The unnatural artifices that we will integrate into our physiology must function in an assigned manner. Invariably will determine based on their purpose, if the function is served well or not. Has it as in Eve's case become obsolete? What is the cost/benefit of a new model? Will it be reliable? This idea is no different from evaluations of value we make when we purchase a car or a television. In our future dependence of in-the-body technology, we will become living appliances, evaluating ourselves the way we evaluate ovens, computers, and automobiles. Will we remain part of nature in the way we are part of nature now?

U.S. Supreme Court Justice Burger famously wrote, nature consists of all things, except "... anything under the sun that is made by man." Under the sweep of this statement, in what category do we place someone who is part nature and part "made by the hand of man?"

To appreciate why in-the-body enhancements may fundamentally change us let us inquire what makes us in part the natural world, and let us determine what makes us in part socially constructed. First, we live in a world and perhaps a universe that behaves in accordance with quantum mechanics—the most widely accepted theory of physics to date. Organized, yet chaotic fields of force pervade all things. Humans as expressions of atoms, molecules, and higher order materiality experience these so-called fields of force in both a biological form and a social state of affairs.

The physical world organizes itself into larger units we call systems. From a phenomenological viewpoint, systems are sets of causal associations where we ascribe intentions, desires, motives, purposes, cause and effect. We hear music, we see something called color, we taste the bitter and the sweet, but at a deeper level, the physical world merely obeys the laws of physics—such as is postulated by the fundamental laws having to do with conservation of energy and entropy. To nature, music is the mere molecular compression of waves; color the particular packet size of energy.

Scientists have usefully bifurcated natural systems according to certain properties of the elements from which they are composed: inorganic and organic.[185] Systems composed of either type of element behave according to fundamental physical laws, but some of these we call life-forms, and they behave according to observed laws of evolution. The observable features include combinations of morphology, physical properties (density, electrochemical, and thermodynamic states) and in some rare instance, consciousness. Consciousness produces the observable effect of intentionality and the ability to think abstractly about such things as human emotion, virtue, and our place in the universe.

Through the exercise of intentionality, we create non-physical systems for our welfare, security, and propagation. The civilizations we create embody communities, governance, legal systems, medical access, economic exchanges, and inventions to make progress commensurate with an evolutionary imperative. We create social systems for human purposes such as survival, altruism, power, affiliation, and achievement. Unlike molecules and metals in fields of force, which are unaffected by human agency, social systems are matters under the direct influence of human intentionality. And in most respects social systems are represented by incorporeal, intangible, states of affairs.

But how does the social reality we construct influence what we come to believe about the nature of ourselves? Humans, through intentionality, assign meaning or status to objects and conditions as they perceive them. There is nothing with which we come into contact that we do not attempt to explain, assign meaning to, or rationalize. And to the extent we use an artifact or idea, we implicitly supply it significance or meaning in regards to its use (trash cans are for disposal, cars for transportation, genomes for specie-specific propagation, mathematics for computation, in-the-body prosthetics to extend life). The definition or the meaning has an associated rule or rule set that supplies the logic for rationally how to identify what an artifact or idea is. Finally, the rule set must be accepted by the community or, more formally, must fall within the ambit of our collective intentionality. But in a posthuman world, collective intentionality will change, formed not by other humans, as constituted today, but by posthumans operating within a network, each separately contributing and adapting to incoming information, each under the influence of rule generating, teme-like, processors embedded in the body. In some sense memes operate this way, transferring information about language or the culture, using the brain directly; whereas temes operate through a transformer, one designed by some hardware engineer or software agent, for reasons that would or would not have been selected had one been free to choose on one's own volition.

For as far back as human history draws us, we identify with our predecessors. We are one with them anatomically, intellectually, and spiritually. The measurable difference is one of progress, where we have filled the psychological and spiritual vessel referred to as our brain. It is there in the construct of the mind that we are conscious of information, enlightened information about a great many things. But emotionally, mentally and anatomically, we are the same people from whom we inherited and now propagate this long line of DNA. What makes Adam and Eve different, (anatomically, emotionally and cognitively different), having been co-opted and altered by human inventions that by their very nature program some feature of a new social reality, a new human potential, a new sense of self, and a new sense of others, some of whom may be technologically inferior or superior.

What is great in man is that he is a bridge and not an end.—
Nietzsche. *Thus Spake Zarathustra.*

FUNCTIONS AND ASSIGNED PURPOSE

The invention of the Internet—with its accessible
databases connected to millions of people—has resulted in
a form of mechanical-electrical intelligence. Through semi-
autonomous search engines, we share a range of socio-
psychological norms, knowledge, and human practices. We
cross fertilize ideas among people from innumerable
disciplines and from around the world, what Dawkin's
coined as a meme.[186] Memes seek to be copied and used
humans as a medium for replicating and exchanging
information. Meme-like machines may not develop
consciousness and self-awareness, but it is likely that by
2085, meme-like artifacts will share the planet with teme-
like machines. We may go about our business as if
operating as autonomous agents, but, in reality, programs
operating in a teme-like sphere of influence will be in
control of what was once considered our inviolable
autonomy.

Through networks, similar to the Internet, we might
imagine that the transferred and retransferred information
communicated via teme-like replication processes, will
create a form of distributed intelligence, called an
infomorph, which will reside in the consciousness of
millions of individuals throughout the world.[187] Although
the technology used for this application may be that of
neural prostheses, where a complete brain is stored in a
prosthetic substrate for whole "mind-transfers"—these
teme-like neural prosthesis will be used to essentially
receive and transmit information, mind to network,
network to mind.[188]

The last time we looked in on Eve, we learned that she
was returning to the work force. We now find her sitting in
front of a full wall screen in her living room, through which
an interview will take place with World Commodities
Trading Corporation. In 2084, few individuals actually
travel to an office, as everything that needs to get done—
especially in commodities trading—is done via an audio-
visual extranet hook up. Prior to the interview, her

prospective employer asked for, and was granted permission to access, the hardware systems Eve had incorporated into her body. They are mainly interested in her cerebral processor with a teme-like neural prosthetic add-on, and whether the version 2.1 will have a decided effect on her ability to do the job. Most cerebral processors are limited and do not substitute for raw IQ, education, and experience, but over the course of time, the enhancements may affect the rapidity and quality of decisions. After all, a competitor with the same basic qualifications as Eve, but who can access information a few milliseconds quicker would perform significantly better over the course of time. The interview is conducted through a series of questions related not only to her experience, but to trading skills, and are all posed by a commercial avatar, an in-silica artifact that looks and acts much like a human. In some ways it is the natural extension of the familiar phone prompts of 80 years earlier. Eve reaches a part of the interview that simulates a series of trading activities to assess performance. The test proceeds for the better part of an hour before Ms. Anderson, a live person from human resources appears on the screen. She is quite congenial and makes small talk before informing Eve that her Version 2.1 processor would need to be upgraded before she could be considered for the job. She extends an offer to return for a follow up interview if and when she upgrades.

That night she and Adam discuss their options. Obtaining an upgrade would improve their income by about twenty-percent, but it would also cost as much as a new car, even after taking into account a ten-percent discount, if she licenses a version 5.0, teme-like neural prosthetic, from a company interested in tapping into her experiences for a research project in rapid technology development. They calculate the payback and decide that in four years they would break-even. At that time, there may be another, more advanced processor and if she were pressed by the employer, she might have to face another upgrade and another investment. In the end, they decide they have little choice but to keep up with the change. If they do nothing, eventually their lifestyle and economic class standing will fall uncomfortably behind their contemporaries.

Dependence on in-the-body technology will change us from completely natural beings, dependent on our natural qualifications, to quasi-natural beings dependent on social constructs formed from in-the-body technology. Through the collective beliefs of people and institutions, a society assigns status through utility, purpose, and worth. Richard Dawkins writes: "Show us almost any object or process, and it is hard for us to resist the 'Why' question—the 'what is it for?' question."[189] A cat serves as a pet; an employee serves as an income generator. As individuals are identified with regard to in-the-body technology and all that may imply, such as to enhance or not enhance qualities, they too will be classified or objectified when asked: "Why?" Or, "what purpose or function does this or that enhancement serve?"

Life as it is presently constituted, which I will refer to as version 0.0, will in the foreseeable future live alongside posthumans. To the unaided eye posthumans will presumably have the same outer appearance as present regardless of how far the internal mechanizations veer from the current norm. This premise is qualified by an assumption that physical enhancement will not produce Goliaths and the common observation that subcultures already adopt styles and mannerisms that often set them apart from the dominant culture. However, once a human has been infused with a life-altering technology beyond some limit, which I will refer to as posthuman version 2.0, they will acquire an assignment of status and an objective valuation in the institutional framework of the assignor.

For the purpose of this discussion, what construct will posthumans represent? Will there be a long transition, during which there is a superior model—perhaps a pre-posthuman version 0.0 that transitions to a posthuman version 2.0? And will society or the institutions that drive technological progress assign a meaning to them and to such an assignment, qualifications leading to an assigned function in society? For example, suppose one were a version 2.0 and expected to live twice the life of version 0.0? And what if the enhancement provided a mental agility or a visual acuity that version 0.0 did not possess? Do these qualities elevate version 2.0 to a more prominent role in society? Is that analogous to what happens in the more highly formally educated and liberal societies today? As humans transition from version 0.0 to version 2.0, will there be any better perception that an evolutionary change is underway, any more than we can look at someone today who has had a heart replaced by a machine, or an athlete on steroids, or a person on Prozac—and say that person is different from the rest of us in some way?

This idea of assigned function creeps into our idea of what things represent and what purpose they serve. Much of how we define a technology—such as a car, computer, drug, or cell phone—has to do with messaging from the purveyors of the technology. In either case, "*Homo sapiens* is a deeply purpose-ridden species", according to Dawkins.[190] For example, I watch cars below my apartment traveling down a well-maintained black-top road. The function I assign to the road is to order vehicular traffic. From another direction, a seagull comes into view. It carries a hard-shelled mussel. Roughly twenty-five feet above the surface, it lets it go. The mussel crashes onto the pavement and its shell cracks open. The obvious function that the road serves the seagull is to facilitate its satisfaction of hunger. The function served by the surface differs based upon the respective needs of those who choose to exploit its properties. In this instance, the function served is not intrinsic, as it is for the molecules that hold the macadam surface of the pavement rigidly in place. The assignment of function, as illustrated by the example, depends upon the manner a human, an animal, or an insect exploits an artifact or a process based upon one's desires, wants, and needs. For animals this may be rather an objective assessment, but human desires, wants, and needs are often defined by the culture and through the pressures applied by its institutions. Function ultimately invokes the notion of use and its cognates such as "utility." I may employ an in-the-body artifact for one purpose, such as to improve my mental agility, but an employer may look at it as an income generator, not very different from how they might look at an upgrade to their telephone system.

Mountains, molecules, metals, and mammals (i.e., natural occurrences) carry out causal processes independent of desires, wants, and needs. Said another way, natural processes precede human purpose or function. Life exists as a mere host for DNA; it is vehicle for DNA survival into perpetuity. [191] Life was believed monopolized by DNA's persistence objective of limitless replication, before the recognition of memes and temes, both of which also apparently use life to further limitless replication. For example, a tree does not grow to accommodate human purpose, it simply grows. A child may see it as a challenge to climb. A farmer may see it as bearing fruit to sell. I may believe that it serves to function as shade for my home and thereby assists in reducing the load on my air conditioner. But the tree's essence, manifest through its DNA's mechanism to survive, has little to do with intentions, certainly not with human intentions. This is rarely the case with artifacts of creation for which humans seek answers to "why" or "for what purpose."

Humans do not exist to accommodate human or institutional purposes. It would seem odd indeed to say that life comes into being to serve a purpose—whether commercial, political, or to artificially enhance life—since the very notion alters what it means to be human. This will change in the posthuman era when an institutional reality will be based on "enframing" or the ultimate objectification and technological manipulation of the human form, precisely because it will not be in large part regarded as a naturally occurring process.

...Heidegger began to distinguish modern "subjectivism" (the modern subject's quest to completely control the objective world) from "enframing" (the objectification of that subject whereby everything gets reduced to the status of an intrinsically-meaningless "resource," Bestand, merely standing by to be optimized and efficiently ordered for further use).[192]

The optimum function desired to be achieved by a thing or process will imply its purpose. For example, as we reach the posthuman era, the function of a designer gene, a drug enhancer, or an in-the-body computer will be less motivated to improve health or enhance the individual, as it will be to create a super society. According to Iain Thomson, Heidegger understood such technological optimization,

> In the late 1930s,... Heidegger seems first to have recognized this objectification of the subject in the Nazis' coldly calculating eugenics programs for "breeding" a master race, but (as he predicted) that underlying impulse to objectively master the human subject continues unabated in more scientifically plausible and less overtly horrifying forms of contemporary genetic engineering.[193]

Who should speak to the moral issues of our time? Who ought to decide what biological forms we create by the "hand of man that would be morally acceptable?" And, whether purpose is a moral construct in speaking about the human subject? I find it especially incumbent on scientists, sociologist and ethicists, to account for the full range of moral risks that will emerge dependent on the differing forms of in-the-body-technology.

While atoms are eternal, the objects compounded out of them are not.—Democritus, Greek Philosopher, 5th Century.

FORM IN CYBORGIZATION

It would be pure fancy to imagine that I could swim across the Pacific, but for a whale, true to its form, this possibility exists. A whale derives the ability to swim vast distances from its endowed form. To the degree living things exploit the full measure of their form, they stand to fulfill their teleonomic ends.[194] This does not imply that evolution itself has such a purpose. Huxley pointed out that, "The ordinary man, or at least the ordinary poet, philosopher and theologian, always was anxious to find purpose in the evolutionary process. I believe this reasoning to be totally false."[195] But what happens when the "form" survives by evolving, not in response to Darwinian selection and adaption, but from human design, a program that unleashes a non-natural process at a cytological level, where even the chromosome finds no sanctuary?[196] Ervin Laszlo, a pioneer in system's thinking observed that:

> The living organism keeps itself in running condition as long as it can, and performs repairs if it gets damaged (these are the processes of healing and regeneration). But very complex organization organisms are unable to keep this up indefinitely, and succumb to internal exhaustion even when relatively undamaged (the process of aging). To survive, such species have managed to develop a way to perpetuate themselves by a form of super repair: reproduction. Instead of replacing a damaged or worn-out part, they replaced the entire organism.[197]

In 1880, Walter Flemming named certain constituents in the central part of the cell "chromatin" after the Greek work for "color." He noticed that during cell division the chromatin collects into thread-like pairs, which were later termed "chromosomes." In the early 1900s, Walter Sutton, and independently Theodor Boveri, found that chromosomes were paired, and they conjectured that chromosomes may be the carriers of Mendel's heredity "factors"—which were later referred to as genes.[198] After observing chromosomal movements during meiosis, (where the cell divides and differentiates) he advanced a chromosomal theory of heredity. [199]Biologists now know that chromosomes are the seat of a complex organization where errors, repairs and now modifications at the chromosomal level take on evolutionary significance. Fast forward a hundred years to 2008, when Craig Venter addressed an audience at a TED conference:

... But we wanted to go much larger: we wanted to build the entire bacterial chromosome—it's over 580,000 letters of genetic code—so we thought we'd build them in cassettes the size of the viruses so we could actually vary the cassettes to understand what the actual components of a living cell are. Design is critical, and if you're starting with digital information in the computer, that digital information has to be really accurate.... Part of the design is designing pieces that are 50 letters long that have to overlap with all the other 50-letter pieces to build smaller subunits we have to design so they can go together. We design unique elements into this....This is a major mechanism of evolution right here. We find all kinds of species that have taken up a second chromosome or a third one from somewhere, adding thousands of new traits in a second to that species. So, people who think of evolution as just one gene changing at a time have missed much of biology. There are enzymes called restriction enzymes that actually digest DNA. The chromosome that was in the cell doesn't have one; the chromosome we put in does. It got expressed and it recognized the other chromosome as foreign material, chewed it up, and so we ended up just with a cell with the new chromosome.... And with a very short period of time, all the characteristics of one species were lost and it converted totally into the new species based on the new software that we put in the cell. All the proteins changed, the membranes changed; when we read the genetic code, it's exactly what we had transferred in. So, this may sound like genomic alchemy, but we can, by moving the software of DNA around, change things quite dramatically. Now I've argued this is not genesis; this is building on three and a half billion years of evolution.[200]

In our quest for a satisfactory life within the confines of our unique form, we each take different paths making choices along the way, each of us driven by notions of "what is good" or "what is worth living or dying for."[201] Perhaps for most individuals "the good" means finding happiness.[202] It is not the intention here to engage in a discussion as to "what is good" specifically, but to observe generally that the content of what we strive to achieve, purpose if you will, takes place in the framework of a psychological, physical and societal architecture or "form", a human form to be specific. Combined our purpose and form constitutes a large part of who we are, our potentialities and limitations.[203] In this context, individuals have a complete awareness of who they are. But, will the incorporation of in-the-body technology of the magnitude suggested by Venter, whether to extend life, improve mental and physical powers, or more fully integrate (e.g., electronically) into the social/commercial world, change, distort or produce a new emergent posthuman form that creates a new psychological, physical and societal architecture defining a new version of "who we are?"[204]

In the 1940s Julian Huxley, asserted that "now and again there is a sudden rapid passage to a totally new and more comprehensive type of order or organization, with quite new emergent properties, and involving quite new methods of further evolution." What if we managed, as Laszlo says, to develop a way to perpetuate ourselves by a form of super repair, and added a chromosome, perhaps as Venter proposes, to the already existing forty-six we inherited from our parents? Would this not replace, not a damaged or worn-out part, but replace the entire organism, as Laszlo suggests and would this not cause the "totally new and more comprehensive type of order or organization" that Huxley may have had in mind? More than one respected commentator has proposed this idea.[205]

It is likely that new chromosome would contain genes manufactured from synthetic DNA, that is, an intelligent designer gene constructed from the same rootstock as the nucleotide bases of our present genes.[206] These would likely be germline interventions that would add artificial chromosomes to for example the zygote's original forty-six. [207] Because it is a germline accumulation, it would pass on indefinitely to the offspring, changing *Homo sapiens* in the succession line for the foreseeable future. But, the design, its order and organization, like a self-organizing computer program folded into its own DNA hardware, would be based on programming innovation.

The function, structure and processes of cytological systems are dependent on information instantiated in the codons that control the biological process of reproduction, cell differentiation and manufacture of protein. In the present form of DNA, this is self-ordered, organized by processes that are "purposeful" in the teleonomic sense. The process is one of ontogenesis, where genes preprogram the embryo, to follow its instructions to the letter. However the process of phylogenesis, creatively advances the species, one generation to the next, the self-transformation of the entire species.[208] The intelligent designer incorporates the current state-of-the-art for reducing disease, increasing happiness, adding phenotypic features that improve self-image, increasing our mental agility, and the golden chalice, increased lifespans. To what degree do these innovations influence the features of what we regard as the most prominent of our identity, our personalities: openness, extraversion, neuroticism, agreeableness, and conscientiousness? In truth, we can only speculate, but unquestionably the structural change, the change in form, introduces another level of complexity. Let us consider how this complexity might work.

Corning elaborates on what Huxley meant by emergent in the context of complexity and what I mean by self-organizing computers and their software, since after all, the designed gene might reasonably be considered as such:

Complexity... is an emergent phenomenon... Emergence is what "self-organizing" processes produce... the reason why there are hurricanes, and ecosystems, and *complex organisms like humankind...* It was originally coined during an earlier upsurge of interest in the evolution of wholes, or, more precisely, what was viewed unabashedly in those days as a "progressive" trend in evolution toward new levels of organization culminating in mental phenomena and the human mind. This long-ago episode, part of the early history of evolutionary theory, is not well known today or at least not fully appreciated.[209]

Hundreds of new bioengineered products are abandoned somewhere between the laboratory and the marketplace. In most modern societies protocols require rigorous testing of the efficacy and safety of new drugs, prosthetics and treatments. In the future, there will be thousands of one-of-a-kind, half-baked engineering models tested and scrapped, before two or more nearly complete prototypes are created. There will be pre-production runs before the technology designated for a test group proceeds. It is likely that great opposition will mark any initiative to add a new chromosome to the gene pool. However, as with opposition to genetically modified food, in vitro fertilization, and stem cells for therapeutics and cloning, there will come a day when the culture accepts the idea that the benefits of a life extended indefinitely, will outweigh the initial repugnance. Part of the acceptance will come from assurances that it will be only through years of testing— first in simulated computer models and animals, then on volunteers, then on test populations—that the procedure will have been deemed safe enough for mass consumption.

Computer programs have long been able to model the behavior observed in the natural sciences. Without this ability, scientists and engineers could not predict the weather, population growth, or how a prosthetic might work before it's actually made. Currently, a wave of activity in bio-informatics touches upon information theory, cellular communication, the function of enzymes, control mechanisms, and feedback systems relevant to the genome. Bioethicist James Hughes suggests that a responsible course for the ultimate genetic manipulation of humans would be to build computer models of the human genome and the proteins for which it codes. And as in most other engineering programs, the model would be expected to eventually lead to safe genetic modifications on humans.[210]

By the time chromosome insertion reaches a safe medical status, society already will have had years of experience with biological, (albeit non-chromosomal) prosthetics. These would not add a chromosome, but would only add to the existing genome—first for therapeutic necessity and then for elective enhancements. The fear factor will be low, as has proven to be the case in electronic, mechanical prosthetics or organ transplants. Chromosomal addition, when the time comes, will be neither more revolting nor revolutionary than DeBakey's first mechanical heart.[211] Through modeling and testing, biomedical companies will assure the government regulators of efficacy and safety. The consumer will satisfy its aim of bequeathing offspring more permanent genetic enhancements. By this time they will already have widely accepted the old-style mechanical/electronic implantable enhancers that had been installed post-birth. These enhancements will be different, because these changes go straight to the germline, where DNA is for all practical purposes, immortal, and in this way able to passed down the enhancements from generation to generation.

After the fact could one say that the post-existing form corresponded to the pre-existing form in any meaningful way? Does this change the game putting us on a one-way track, both humanly apocalyptic and "cyborgically dystopic"?[212]

In human enhancement literature, the most frequently discussed topics for deep-seated enhancements—such as a new chromosome might provide—concern extending lives and increasing intelligence.[213] There are fewer questions as to "how" this will occur, as questions about what such enhancement might accomplish. But even agelessness may not be desirable unless we were assured it would be a status, which halted the progression of age at some preferred age—such as at forty. To be stuck at the ripe old age of 200 for the next millennium would not be particularly attractive to most people. Likewise, "intelligence enhancement" means exactly what? Analytical, artistic, book smart, simply wise, or all the above? Would all of the above even be an option?[214] Would economics deprive the masses of equal access?

Verner Vinge, when interviewed by *Wired* magazine, expressed a not uncommon futurist vision:

> *Wired*: "What are some of the scenarios for how the Singularity might unfold?"
> Vinge: "I think there are all sorts of different paths to the Singularity, at least five pretty different paths... in many ways is the most attractive... is the notion of "intelligence amplification," which is that we get user interfaces with computers that are so transparent to us that it's like the computer is what David Brin calls our 'neo-neocortex.'What's nice about that is that we actually get to be direct participants, and in that particular case, when I say that the post-Singularity world is unintelligible, well, yeah, it is unintelligible to the likes of you and me, but it would not be unintelligible to the participants that are using intelligence amplification."[215]

Unlike other animals that inhabit the planet, humans have the power to revise nature, e.g., to countermand human suffering. In the matter of enhancing life forms, it is human *form* from which *human essence* flowers to serve our species. If we affect form, we run the risk of affecting human essence.[216]

The ontology side of *essence* deals with the predicate for its mode of existence. When we refer to its absolute form, we might consider features that are independent of what anyone thinks—that is, by removing all cultural considerations. The existence of the genetic code is not subjective. It is objective because it does not depend on human perception. Essentially, we alone insert it in the category of "living things." However, despite our proclivity to define and categorize the arrangement of DNA molecules, it has always been and remains observer independent, intrinsic to nature. The fact about DNA, for instance, is that it provides the plans that manufacture protein essential to life as we know it. This fact is indisputable and epistemically objective—a fact preceding human existence.

Ontologically we exist according to our identities, one which is numeric and the other narrative. David DeGrazia of George Washington University sees one blending into the other:

> ... our numerical identity consists (at least partly) in some sort of psychological continuity. In most psychological theories (citations omitted), the relevant type of continuity is a continuity of experiential contents, or the maintaining of psychological connections, over time. Examples... are having an experience and later remembering it, forming an intention and later acting on it, and the persistence of certain beliefs, desires, and character traits. In some psychological theories (citations omitted), by contrast, the relevant type of continuity is the continuation of one or more basic psychological capacities, which may remain despite loss of memories and other experiential contents. Examples include a basic capacity for reasoning or, even more minimally, the capacity for conscious experience; either capacity could survive complete amnesia. On any psychological theory, permanent loss of the capacity for consciousness—and, therefore, of any specific types of mental states or psychological connections—entails the end of our existence.[217]

At this moment, my numerical identity is: "I shall remain who I am throughout my life, regardless of my age or appearance." Those who know me identify me by name because they perceive me in some invariant way. What essential feature about me does not change so that I remain who I am? Features change slowly, but my personality changes even slower, beneath this there may be some tic, or simply a manner of frowning or smiling, combined to impart uniqueness—called my identity. A mirror might serve as a metaphor for imagining the relationship between our human essence and our psychological self. The reflections we see are not part of the mirror, but are perceived by us because of the mirror. The mirror has material substance, but its reflections are mere forms of disembodied energy, reassembled somewhere within the interstices of my brain. The reflections observed are phenomenological. In my metaphor the mirror remains indispensable to the production of the conscious reflections of my essence. We cannot physically possess our essence because it is intrinsic to our being. By further analogy, *water does not contain hydrogen and oxygen—it is hydrogen and oxygen.* The DNA code exists not as something apart from that which we are; it is that which we are. We possess our essence differently than we possess artifacts within our environment—things that we use as extensions of ourselves, things we invent and install as adjuncts into our anatomical selves.

That said, we can imagine that when we change elements affecting genotypes, phenotypes, or natural behaviors, we run the risk of changing our present essence into a new one, an essence that potentially changes our numerical identity and certainly our narrative identity. Perhaps C.S. Lewis said it best:

In order to understand fully what Man's power over Nature, and therefore the power of some men over other men, really means, we must picture the race extended in time from the date of its emergence to that of its extinction. Each generation exercises power over its successors: and each, in so far as it modifies the environment bequeathed to it and rebels against tradition, resists and limits the power of its predecessors. This modifies the picture which

165

is sometimes painted of a progressive emancipation from tradition and a progressive control of natural processes resulting in a continual increase of human power. In reality, of course, if any one age really attains, by eugenics and scientific education, the power to make its descendants what it pleases, all men who live after it are the patients of that power. They are weaker, not stronger: for though we may have put wonderful machines in their hands we have pre-ordained how they are to use them.[218]

Institutions draft mission statements to remind them about the purposes they serve in society. Profit-making businesses fashion mission statements around language that ultimately reflects survival and profits. Understandably, they protect their commercial interests, and for those in the medical device or pharmacological industries this is done through cutting-edge technology backed by intellectual property rights. Humans have missions or purposes different from corporations, ones not steeped in technology, but that foster a particular kind of life, one based on what Alasdair MacIntyre refers to as the narrative tradition:

> A central thesis then begins to emerge: man is in his actions and practice, as well in his fictions, essentially a story-telling animal. He is not essentially, but becomes through his history, a teller of stories that aspire to truth. But the key question for men is not about their own authorship; I can only answer the question 'What am I to do?' if I can answer the prior question 'Of what story or stories do I find myself a part?' We enter human society, that is, with one or more imputed characters— roles into which we have been drafted —and we have to learn what they are in order to be able to understand how others respond to us and how our responses to them are apt to be construed.[219]

The narrative should not be relegated to non-person interests, such as corporations, or anthropomorphized by technology. Capturing this thought, Margaret Farley writes:

> At the heart of tradition, however, is a conviction that creation is itself revelatory, and knowledge of the requirements of respect for created beings is accessible at least in part to human reason. This is what is at stake in the tradition's understanding of natural law... natural law theory does tell us where to look: that is, to the concrete reality of the world around us, the basic needs and possibilities of human persons in relation to one another, and to the world as a whole.[220]

Given the promise of in-the-body technology, we can only understand its intrinsic value through the precepts of our traditions. We are not dealing with drugs here, but relatively permanent adjuncts to our anatomy. These artifacts will constitute a part of what contributes to the whole, and therefore, whether we sense or are conscious of the thing that keeps our heart beating is irrelevant—it is also who we are. An evaluation on these terms should guide our judgment on whether the very nature of advanced in-the-body technology bars ownership in the monopolistic or corporate sense; in short whether any ownership of the technology in the customary use of the notion can be morally justified.

For if we reach the point where we are oblivious to or have forgotten the unadulterated *Homo sapiens* as they once were, we will then have faced the "ultimate destruction of our entire species." On that dawn, we will have been replaced by a kind of species that has little relationship to what we are today. Nearly seventy years ago, C.S. Lewis warned about Man's final mastery over Nature, and the inevitable drift into a future world where knowledge about what the old world completely vanishes:

> ... Now I take it that when we understand a thing analytically and then dominate and use it for our own convenience, we reduce it to the level of 'Nature' in the sense that we suspend our

judgements of value about it, ignore its final cause (if any), and treat it in terms of quantity. This repression of elements in what would otherwise be our total reaction to it is sometimes very noticeable and even painful: Something has to be overcome before we can cut up a dead man or a live animal in a dissecting room. These objects resist the movement of the mind whereby we thrust them into the world of mere Nature. But in other instances too, a similar price is exacted for our analytical knowledge and manipulative power, even if we have ceased to count it. We do not look at trees either as Dryads or as beautiful objects while we cut them into beams: The first man who did so may have felt the price keenly, and the bleeding trees in Virgil and Spenser may be far-off echoes of that primeval sense of impiety. The stars lost their divinity as astronomy developed, and the Dying God has no place in chemical agriculture. To many, no doubt, this process is simply the gradual discovery that the real world is different from what we expected, and the old opposition to Galileo or to 'body-snatchers' is simply obscurantism.... The great minds know very well that the object, so treated, is an artificial abstraction, that something of its reality has been lost.[221]

Fukuyama reminds us about the reality of who we are at this moment in history: "... the sum of behavior and characteristics that are typical of the human species, arising from genetic rather than environmental factors."[222] Perhaps human rights policy should not embolden those who would change our present essence into a new one, for one essential reason: It tampers with "the way we are", our identity. [223] Public policy should reflect moral ends that a society might rationally adopt. [224] "Form" as discussed here should be written into legislation as a fair, objective point of reference. In respect to U.S. technology, policy ought to aim at securing the integrity of our natural patterns of formation; otherwise, we risk moving into forms that, looking forward, might be regrettable.

CONCLUSION

Debtors[225]
by Jim Harrison

They used to say we're living on borrowed
time but even when young I wondered
who loaned it to us? In 1948 one grandpa
died stretched tight in a misty oxygen tent,
his four sons gathered, his papery hand
grasping mine. Only a week before, we were fishing.
Now the four sons have all run out of borrowed time
while I'm alive wondering whom I owe
for this indisputable gift of existence.
Of course time is running out. It always
has been a creek heading east, the freight
of water with its surprising heaviness
following the slant of the land, its destiny.
What is lovelier than a creek or riverine thicket?
Say it is an unknown benefactor who gave us
birds and Mozart, the mystery of trees and water
and all living things borrowing time.
Would I still love the creek if I lasted forever?

I began with three goals in mind. The first, to explore
current in-the-body technology and that on the near
horizon; second, to show how in-the-body technologies will
spawn changes in our social reality; and finally, to explore
the reasons we should control these technologies of the
future.

Civilization attains just ends when the sum of its
intellectual and technological achievement brings order out
of chaos, reason out of ignorance, and compassion out of
suffering. It attains moral ends when its culture grounds
itself in a fair political structure, a fair economic
distribution, and a sense of respect for all its members—
especially its weakest and most vulnerable. We cannot

169

ignore that we live in a time of great world division: social, religious, political, and moral. The moral divide does not cut across political, religious, or ethnic borders as much as it cuts across sectors within societies that live in moral uncertainty. In the modern world, there is no consensus about manufacturing new life forms. There is no consensus about access to either lifesaving drugs, or cornering information vital to biological scientific progress, or manipulating *Homo sapiens* from their natural state.

The state creates an economic, legal, and social framework that fosters the principles of democracy and free enterprise. In the exercise of these goals, tensions exist due to the opposing forces of individual freedom, competition, and the state's duty to the general welfare. If we were to follow any tension back to its source, we would likely find it anchored in socioeconomic ideologies and moral convictions both religious and secular. In a free society, the people ought to participate in the debate over the present course of cyborgization.

In the new world of bioengineering, private ownership of certain forms of cyborgization property will become central to lifesaving products. We need to reevaluate what implications lie ahead for a society that espouses principles for distributive justice. The inequitable distribution of cyborgized property will affect each of us or our progeny—just as the inequitable distribution of health care, education, clean air, water, water rights, or land does to those that live among us presently.[226] It needs to be decided whether any corporation, institution, or individual should have the right to private ownership of certain forms of cyborgized property—most notably, selected methods of extending life and enhancing intelligence.

In matters affecting commerce, just a few powerful but partisan voices speak for the dominant culture and define for the rest of us what we should regard as the good, the right, and the responsible. Many of the nation's elderly and children go without adequate health care, or they go hungry. These realities cannot be changed through a more or less liberal cyborgization policy to make everyone more intelligent through in-the-body nanotechnology, electronics, or genetic manipulation. Nonetheless, all too often, partisans—moved by the ancient and primitive motivations of self-interest, wealth and power—wage lobbying campaigns in an effort to control such things as health care and the property rights to drugs and in-the-body technologies. We live in an economy where the commodification of goods and services overshadows the greater purpose intended.[227] Voices of the impoverished are never heard.

Far from an obvious consensus exists in regards to allowing unregulated cyborgization research to progress. Therefore regarding the wisdom of patenting such technology, policy makers need to listen to the voices of a wide range of constituents in order to judge in which direction we point the moral compass.

We must keep in mind that no single unified system's theory exists that considers invention on the level of motivating factors, economic effects, and societal goals— including the relatively new ethics of cyborgization. Invention itself embraces a universe of art, economics, psychology, science, and engineering. Before a theory of Keynesian economics existed, we had a fair idea of such concepts as systems of labor, capital, supply, and demand. Such a lucid prefatory structure for cyborgization law has not been established. Part of the problem in achieving this objective is that cyborgization property deals with political rights of the individual and with other fundamental values and prescriptions for society, as mentioned throughout this book. Cyborgization policy needs to consider law, science, economics, ethics and politics. In cases where the technology has profound implications for social justice or mischief, we need to be especially mindful that without a theoretical underpinning, we often make arbitrary judgments that in the end prove counterproductive to humanitarian interests.

In the cyborgization property sphere, diverse influences work their will on nascent policy and legislation. Yet, as the subject deals with economic interests, the wind blows predominantly in the direction of the corporation. Some corporate goals may align with the bodies of common ideals or ideologies.[228] Clearly, the ideologies that represent mainstream American society may not translate well into corporate interests. We see this in regards to environmental, health care, and tax laws. For those that stand to gain (i.e. corporate and other special interests), they move in lock-step to influence legislative offices and the halls of justice. When it finally matures, cyborgization property law will be a subject that raises more questions than can be answered by black-letter law. As such, it can be easily misunderstood and manipulated for the few who stand to make the most of legislation, even when the best interest of citizens point in another direction. After all, politicians and executives are not philosophers, scientists, mathematicians, or technology lawyers.

Our human form should inform our humanity. [229] Until now our specie has lived a cycle of natural patterns that form our existence. That's about to change with life expectancies anticipated to increase significantly during the next thirty to fifty years, due in large part to the internalizing of technology. Science and technology benefit humanity by reducing the sum of suffering, adding quality to life generally, and serving to answer the indomitable quest for more and exacting knowledge about our reality. However, in the hands of the irresponsible, scientific, and technological advance can rip through the fabric of a delicate world and destroy natural patterns—from exacerbating global warming to driving a species extinct from the release of uncontrollable technological scourges. One cannot predict with any accuracy where technology will take us. But utilizing discrete in-the-body implantation on a large scale, for multiple organ remediation, life extension, and human enhancements will change natural human patterns. These patterns do not merely express the rules of reproduction and survival; they express the form of life through a consciousness and a conscience of communal constructs. These constructs include a moral catechism—albeit one authored within the narrative of the individual, her culture and the times to which she is born. If, through an irresponsible and irreversible application of technology, we were to damage the patterns formed by Nature, we would be accountable for affecting the moral ecology upon which all humanity—as we have come to appreciate the term depends. I believe that moral ends supported in sound legal regulation and practice, better secure the integrity of Nature's patterns. This book hopes to bring an educated eye and common sense to a largely unbridled, free enterprise that will markedly alter the state of current affairs. We need to be prepared.

1 Jim Harrison, *Songs of Unreason.* (Copper Canyon Press, 2011). Reprinted with permission.

2 Craig Venter, "On the verge of creating synthetic life," *TED2008,* (March 2008), http://www.ted.com/talks/craig_venter_is_on_the_verge_of_creating_synthetic_life.html (Last viewed, 8/12/2012).

3 "...most scientists hold that the first organisms on Earth were much like bacteria of today...," http://www.actionbioscience.org/newfrontiers/jeffares_poole.html (Last visited, 8/7/2012).

4 "All people today are classified as *Homo sapiens.* Our species began to evolve nearly 200,000 years ago. It is clear that early *Homo sapiens,* or modern humans, did not come after the Neandertals, but were contemporaries. However, it is likely that both modern humans and Neandertals descended from *Homo heidelbergensis.*" http://anthro.palomar.edu/homo2/mod_homo_4.htm(Last visited, (8/7/2012).

5 Robert R. McCrae and Paul T. Costa, Jr. "Validation of the Five-Factor Model of Personality Across Instruments and Observers," *Journal of Personality and Social Psychology*, Vol. 52, No. 1 (1987).

6 E.g., global warming, vast metropolises, breakthrough advances in travel, terrestrial, air and interplanetary space.

7 The IBM Watson computer competed with top performing guests on the game show Jeopardy out-scoring them in an entertaining back and forth that included natural speech responses. The computer has the ability to "understand" the context of the questions, and search a vast data base that includes natural language documents as well as such standard databases such as Wikipedia and thousands of library references.

8 The word *prosthetic* is derived from the Greek word *prósthesis* meaning addition, application, or attachment which in modern times refers to artificial body parts.

9 John Locke considered personal identity to be founded on consciousness, and not on the substance of either the soul or the body. John Locke, *On Identity and Diversity in An Essay Concerning Human Understanding,* Book II Chapter XXVII (1689).

10 Charles Darwin, *On the Origin of Species by Means of Natural Selection, or the Preservation of Favoured Races in the Struggle for Life, (1859)* (Penguin, Baltimore, 1968).

11 Francis Fukuyama, *Our Posthuman Future: Consequences of the Biotechnology Revolution*, (Farrar, Straus & Giroux, 2002). p.152.

12 [cyb(ernetic) + org(anism).] The *American Heritage® Dictionary of the English Language*, Fourth Edition (Houghton Mifflin Company, 2009).

13 Susan Blackmore studies memes: self-replicating "life forms" that spread via human consciousness from brain to brain. She believes that humanity has spawned a new kind of meme, the teme, which spread via technology, that "invent" ways of keeping itself thriving. See, Susan Blackmore: *Memes and "temes" TED2008*, (Jun 2008), http://www.ted.com/talks/susan_blackmore_on_memes_and_temes.html. (Last visited, 10/26/12).

14 Max More, *On Becoming Posthuman*, (1994), http://www.maxmore.com/becoming.htm. (Last visited, 10/26/12).

15 Susan Blackmore: Memes and "temes".

16 Will Weissert, *Associated Press*, (7/14/2004), http://www.msnbc.msn.com/id/5439055/ns/technology_and_science-tech_and_gadgets/t/microchips-implanted-mexican-officials/.

17 "RFID Microchip-First TV Commercial," *Verichip Corp.*, http://sgtreport.com/2012/06/verichip-corp-rfid-microchip-first-official-tv-commercial/ (Last visited 914/2012).

18 Although today's silicon-based microprocessor is as small as 20 microns these processors use 32 nanometer technology, which incorporate six-cores hyper- threaded chips running at roughly 3.3 gigahertz. Since each core executes two threads the theoretically processing speed is a 40 billion flops (flop is one program cycle or operation per second). Before 2020 silicon chips will use 11 nanometers technology incorporating fifty-cores allow 100 simultaneous hyper- thread to execute 330 billion flops.

19 1850s The Dubois chloroform inhaler makes anesthetics safer. 1863 Marey invents the pulsewriter to trace the pulse wave on a moving plate. 1873 Callender performs the first successful heart surgery chloroforming

the man to remove the needle and suture the site. 1880s Lister
revolutionizes surgical practice by introducing aseptic surgery decreasing
surgical deaths. 1887 Waller invents an electrocardiogram using a mercury
column, which pulsated with each heartbeat. 1895 Roentgen, discovers "X-
Rays." allowing doctors to study the heart on fluoroscopes. 1896 Riva-
Rocci invents the sphygomanometer to measure blood pressure. 1899
Williams uses the "roentgenoscope" to demonstrate enlargement of the
heart, aneurysms and pericardial bleeding. 1900s Blundell performs
transfusions on women hemorrhaging from post-partum child birth. 1903
Einthoven invents a string galvanometer electrocardiograph. 1908 First
transfusion using Landsteiner's ABO typing technique. 1911 Curie
discovers radium. 1914 White publishes his findings regarding taking over
27,000 EKG's. 1916 McLean isolates heparin making anticoagulation
possible.1917 Magill invents the endotracheal tube.1923 Cutler, Beck and
Levine operate on an 11 year old girl for mitral stenosis employing a
valvulotome. 1929 Fleming discovers penicillin. 1929 Gibbs develops an
artificial heart with two bellows in a round brass container. 1929 Forssman
develops the technique of cardiac catheterization. Inserts the catheter to
inject opaque dye into his heart to outline the organ's chambers on X-Ray
photographs. 1931 Gibbon, Jr., conceives the idea of the heart-lung machine
for extra-corporeal circulation to remove pulmonary emboli from
moribound patients. 1931 Gibbon and Churchill, first use of phenobarbitone
for anesthesia. 1934 DeBakey invents the DeBakey heart pump. 1935
Gibbon, Jr., uses first successful heart-lung machine. 1939 Drew, reports
the use of plasma is preferred over whole blood for shock, burns and open
wounds. 1941 Cournand performs the first cardiac catheterization on a
human.1940s Blalock and Taussig, co-design the surgical technique for
treatment of pulmonary stenosis, ventricular septal defect, overriding aorta,
right ventricular hypertrophy. 1943 Waksman discovers the antibiotic
streptomycin used in the treatment of tuberculosis and other diseases. 1944
Crawfoord surgically repairs coarctation of the aorta in a human.
http://perfline.com/textbook/local/mvinas_chronol.htm (Last visited
8/7/2012).

20 Mechanical hearts have been under development since the 1930s, when
Dr. Alexis Carrel collaborated with Charles Lindbergh to invent a
mechanical heart. But, nothing of the sort was implanted into humans until
1966, when Dr. Michael DeBakey implanted a booster pump to temporarily
assist the heart. In 1969, Dr. Denton Cooley temporarily implanted a
completely artificial heart. In 1982, Dr. Robert Jarvik implanted the first
permanent artificial heart.

21 Implantable Cardioverter Defibrillator is implanted in patients at risk of
sudden cardiac death due to ventricular tachycardia (faster than normal

rhythm) or ventricular fibrillation (uncoordinated contraction, often spasm) of the cardiac muscle of the ventricles. It includes one or more wires that pass through a vein to the right chambers of the heart and typically attach in the apex of the right ventricle. An electrical impulse generator delivers sufficient energy to defibrillate the heart if it senses that the anti-tachycardia pacing is not bringing the heart back to a normal rhythm or if the unit senses a ventricular fibrillation.

22 "The two basic types of VADs are a left ventricular assist device (LVAD) and a right ventricular assist device (RVAD). If both types are used at the same time, they're called a biventricular assist device (BIVAD). The LVAD is the most common type of VAD. It helps the left ventricle pump blood to the aorta. The aorta is the main artery that carries oxygen-rich blood from your heart to your body." http://www.nhlbi.nih.gov/health/health-topics/topics/vad/ (Last visited 8/8/2012).

23 Rosemary Black, "Former vice president Dick Cheney now has no pulse; Heart pump like artificial heart." *Daily News* (New York) January 5, 2011, http://articles.nydailynews.com/2011-01-05/entertainment/27086458_1_ mechanical-heart-artificial-heart-surgical-director (Last visited 8/7/2012).

24 Before the 1980s prosthetic legs were designed to add a length extension from the residual limb to the ground, but in 1981, the "Seattle Foot" modernized limb prosthetics by introducing the idea of an energy storing prosthetic foot, which performed more akin to a natural organ that stored energy, acting more anatomically spring-like. Pistorius was outfitted with the Flex-Foot Cheetah, based in part on this idea, invented by Van Phillips, an amputee himself.

25 "The word "transhumanism" was first coined in the 1940s by Julian Huxley, a central figure in the evolutionary synthesis movement. It reemerged as an idea in the 1960s and 1970s when colleges were offering courses in future studies and then again in the 1990s laying the intellectual groundwork for futurist philosophy know today as the transhumanist movement. Transhumanists offer analysis and speculate about the degree to which we can and will refine the human species. A central assumption among us is that there's significant potential for the re-engineering of humanity; in modern practice we have scarcely begun to scratch the surface, but our visions of what may be possible in terms of modification and enhancement is startlingly vast." George Dvorsky, "Transhumanism and the Intelligence Principle," *Sentient Developments, science, futurism, life,*

(6/22/2009).
http://www.sentientdevelopments.com/2009/06/transhumanism-and-intelligence_22.html (Last visited, 8/8/2012).

26 Nanotechnology: Structure related to features of nanometer scale (10 meters): thin films, fine particles, chemical synthesis, advanced microlithography, and so forth. Biotechnology: Process and structural technology related to the use of parts or products of living organisms, in natural or modified forms. Information technology: Process and structure related to computer hardware and software, including networking. Cognitive science: The study of intelligent systems, with reference to behavior as computation.

27 Noah Shachtman,"Army Dreams: Super-Strong, Laser-Proof, Genius G.I.s,"(5/4,2009), http://www.wired.com/ dangerroom/ 2009/05/ army-dreams-of-super-soldiers/ (Last visited 8/26/1012).

28 A cyborg is a being with both biological and artificial (e.g. electronic, mechanical or robotic) parts. The term was coined in 1960 when Manfred Clynes and Nathan S. Kline used it in an article about the advantages of self-regulating human-machine systems in outer space. See, "Cyborgs and Space," in Astronautics (September 1960), by Manfred E. Clynes and Nathan S. Kline. D. S. Halacy's Cyborg: Evolution of the Superman in 1965 featured an introduction which spoke of a "new frontier" that was "not merely space, but more profoundly the relationship between 'inner space' to 'outer space' – a bridge...between mind and matter." See, D. S. Halacy, *Cyborg: Evolution of the Superman* (New York: Harper and Row Publishers, 1965), p.7.

29 Max More, *On Becoming Posthuman*, (1994), http://www.maxmore.com/becoming.htm (Last visited, 10/18/2012).

30 The word *robot* was coined by writer Karel Čapek in his play *R.U.R. (Rossum's Universal Robots)*, published in 1920. See, Zunt, Dominik. "Who did actually invent the word "robot" and what does it mean?". The Karel Čapek website. http://capek.misto.cz/english/robot.html. Retrieved 2007-09-11. The play depicts a factory that manufactures artificial people called *robots*, creatures who, due to their form and that they reason, are easily mistaken for humans.

31 http://www.guardian.co.uk/technology/2004/jun/10/onlinesupplement1(Last visited 10/22/2012).

32 http://www.youtube.com/watch?v=wgmraKtx7XI&feature=related (Last visited 10/22/2012).

33 http://inventors.about.com/library/inventors/bledison.htm (Last visited 11/06/2012).

34 Everett M. Rogers, *Diffusion of Innovations*, fifth Ed. (Simon and Schuster, 2003).

35 Jeanne M. Rhea, et al.,"Next Generation Sequencing in the Clinical Moleciular of Cancer, Advantages and Challenges to Clinical Laboratory Implementation," *Medical Laboratory Observer*, Vol. 43, No. 12, (12/2011).

36 Brandon S. Razooky , et al., "Microwell devices with finger-like channels for long-term imaging of HIV-1 expression kinetics in primary human lymphocytes," Lab Chip, 2012, Advance Article (First published on the web Aug 21, 2012). The article discusses a micro-device for loading of non-adherent cells into micro-channels for multiday time-lapse fluorescence microscopy of HIV-1 infected patient-isolated T lymphocytes.

37 For example, doctors will soon implant in a patient's brain microarrays with thousands of genes inscribed through etching onto glass chips. It will be a short step to electronically acquire and communicate the data in the microarray directly to a remote central location for analysis.

38 "Information Technology and Cognitive Science," Mihail C. Roco and William Sims Bainbridge, eds.(Arlington, VA: *National Science Foundation*, 2002.) p.300.

39 For well over a decade a continuous glucose monitoring system continuously records interstitial glucose levels in persons with diabetes mellitus. This information supplements, blood glucose information obtained using standard home glucose monitoring devices and may be downloaded and displayed on a remote computer where it may be reviewed by health care professionals. The information allows for identification of patterns of glucose level excursions above or below the desired range and facilitates therapy adjustments to minimize these excursions.

40 Defective p53 allow abnormal cells to proliferate, resulting in cancers—fifty percent of all human tumors contain p53 mutants—are detectable by electronic sensors. See, http://www.bioinformatics.org/p53/introduction.html (Last visited, 11/07/2012).

41 By 2025 there will have been considerable progress in the creation of drugs personalized on the basis of a person's specific genome. Already, on a non-personalized basis, we are reaping the benefits of nano-sized carbon-based buckyballs being employed to "interrupt" the allergy/immune response by inhibiting mast cells from releasing histamines by binding to free radicals more effectively than anti-oxidants currently available.

42 http://www.fountainmagazine.com/Issue/detail/Quantum-Worlds-from-Entanglement-to-Telepathy (Last visited 11/06/2012).

43 J.D. Watson, F.H.C. Crick, "Molecular Structure of Nucleic Acids, A Structure for Deoxyribose Nucleic Acid," *Nature*, (1953).

44 ,Jeffrey L.Bada , *"Stanley Miller's 70th Birthday"*, *Origins of Life and Evolution of the Biosphere,* (Netherlands: Kluwer Academic Publishers, 2000).

45 Utility patents, design patents and plant patents all have varying restrictions on what can be patented. Under 35 U.S.C. 101, the law requires that the subject matter be a "useful" invention. The phrase "anything under the sun that is made by man" is limited by the text of the statute, meaning that one may only patent something that is a machine, manufacture, composition of matter or a process.

46 A full account of the legal case is presented in Daniel J. Kevles, *Ananda Chakrabarty wins a patent: Biotechnology, law, and society, 1972-1980*, Historical Studies in the Physical and Biological Sciences 111-135 (1994). Other Miller–Urey-type electric discharge experiments related to the origin of life were reported by *The New York Times* (March 8, 1953), "Looking Back Two Billion Years" describing the work of William Wollman at Ohio State University, before the Miller Science paper was published in May 1953.

47 See *Parker v. Flook*, 437 U.S. 584, 98 S.Ct. 2522, 57 L.Ed.2d 451 (1978); *Gottschalk v. Benson*, 409 U.S. 63, 67, 93 S.Ct. 253, 255, 34 L.Ed.2d 273 (1972); *Funk Brothers Seed Co. v. Kalo Inoculant Co.*, 333 U.S. 127, 130, 68 S.Ct. 440, 441, 92 L.Ed. 588 (1948); *O'Reilly v. Morse*, 15 How. 62, 112-121, 14 L.Ed. 601 (1854); *Le Roy v. Tatham*, 14 How. 156, 175, 14 L.Ed. 367 (1853).

48 In 1985, Charles B. Shoemaker using recombinant DNA techniques cloned erythropoietin. The subsequent patent relates to the production of a

erythropoietin-type substances that maintain the number of circulating erythrocytes (red blood cells) at a level for delivery of oxygen to the body tissues prepared. The therapeutic use of these substances plays an important part in the treatment of anemic conditions. See U.S. Patent 4,835,260.

49 *Ex parte Allen* 2 USPQ 2d 1425 (1987).

50 U.S. Patent No. 4,736,866, *Transgenic non-human mammals* (issued Apr. 12, 1988).

51 See U.S. Patent No. 6,030,833, *Transgenic swine and swine cells having human HLA genes* (issued Feb. 29, 2000); See U.S. Patent No. 5,972,703, *Bone precursor cells: compositions and methods* (issued Oct. 26, 1999); U.S. Patent No. 6,353,150, *Chimeric mammals with human hematopoietic cells* (issued Mar. 5, 2002).

52 Lee M. Silver, *Remaking Eden: Cloning and Beyond in a Brave New World* (Avon Books, 1997); Allen Buchanan et al., *From Chance to Choice: Genetics and Justice* (Cambridge University Press, 2002).

53 J. Craig Venter et al., "Creation of a Bacterial Cell Controlled by a Chemically Synthesized Genome", *Science*, July 2, 2010.

54 Throughout the Eighteen the and first half of the Twentieth Century, physicians concentrated on antiseptic practices, improving surgical procedures, reducing morbidly due to shock.

55 "Scientists create 'artificial life' – synthetic DNA that can self-replicate," http://io9.com/5543843/ www.sciencemag.org, (Last visited 1/28/12).

56 Craig Venter is a biologist and entrepreneur instrumental in sequencing the human genome and more recently for his role in creating the first cell with a synthetic genome in 2010 and working to construct synthetic biological organisms.

57 Venter, "On the verge of creating synthetic life."

58 "Cramming more components onto integrated circuits," *Electronics Magazine*. p. 4. ftp://download.intel.com/museum/Moores_Law/Articles-Press_Releases/Gordon_Moore_1965_Article.pdf. Retrieved 2006-11-11.

59 Joel Garreau, *Radical Evolution: The Promise and Peril of Enhancing Our Minds, Our Bodies -- and What It Means to Be Human*, (Doubleday, 2005).

60 H. G. Noller, "The Heidelberg Capsule Used For the Diagnosis of Peptic Diseases," *Aerospace Medicine*, (Feb. 1964), pp. 115-117.

61 U.S. pat. 5,279,607.

62 Michael S. Okun, M.D., et al., write: "DBS has provided dramatic improvements in quality of life for patients with PD, tremor, dystonia, and other movement and basal ganglia related brain disorders. As the technology is refined we will learn to improve our treatment of "motor (tremor, stiffness, slowness, balance, gait)," as well as "non-motor (mood, cognitive, and behavioral)" symptoms, perhaps in combination with other therapies." See, http://mdc.mbi.ufl.edu/surgery/am-i-a-candidate-for-deep-brain-stimulation-intro/what-is-the-future-for-deep-brain-stimulation, (Last Visited 11/07/2012).

63 Parylene is a conformal protective polymer coating material utilized to uniformly protect any component configuration on such diverse substrates as metal, glass, paper, resin, plastic, ceramic, ferrite and silicon. It conforms to virtually any shape, including sharp edges, crevices, points; or flat and exposed internal surfaces. It is applied at the molecular level by a vacuum deposition process at thicknesses from 0.100 to 75 microns. No catalysts, solvents or foreign substances are introduced that could degrade the coated surface.

64 The 1938 *Food, Drug, and Cosmetic Act* and its *1976 Medical Device Amendments* give the Food and Drug Administration ("FDA") jurisdiction over medical devices and provide a framework with which the FDA can identify, classify, and regulate medical devices.

65 Medtronics, Inc., developed the IsoMed Constant-Flow Infusion System for bodily implantation that provides intrathecal infusion of morphine sulfate solution in the treatment of chronic intractable pain, or the intravascular infusion of floxuridine for the treatment of primary or metastatic cancer.

66 Pierre Teilhard de Chardin, *The Phenomenon of Man*, (Harper & Row, 1961), p. 244.

67 Combination products merge product types and blur the historical lines of separation between medical products regulated in part by the Center for Biologics Evaluation and Research (CBER), the Center for Drug Evaluation and Research (CDER), and the Center for Devices and Radiological Health (CDRH). Based on the FDA's definition the term combination product includes products having two or more regulated components, i.e., drug/device, biologic/device, drug/biologic, or drug/device/biologic, that are physically, chemically, or otherwise combined or mixed, yet produced and functioning as a single entity.

68 Rudolf von Koelliker and Heinrich Müller in 1856 were the first to discover, that the heart generated electricity. G.E. Burch, N.P. DePasquale. "A history of electrocardiography," *Chicago: Year Book Medical Publishers* (1964).

69 Walter Gaskell, in 1886, demonstrated muscle fibers joining the atria and ventricles when cut caused "heart block" and found that the sinus venosus was the area of first excitation of the heart. *Pub. American Heart Association.* http://circ.ahajournals.org/ (Last visited 8/23/2012).

70 S. Furman, et al., "Reconstruction of Hyman's second pacemaker," *Pacing Clin. Electrophysiol*, (May2005) pp 446-453.

71 An exemplary in-home monitoring system is the LATITUDE.RTM.patient management system, available from Boston Scientific Corporation, Natick, Mass. Aspects of the in-home monitoring system are described in U.S. Pat. No. 6,978,182.

72 Stephen Oesterle, *TEDMED* (2010), http://www.youtube.com/watch?v=ZiQJIpd2n8k (Last visited, 10/21/2012).

73 Some refer to these drugs as smart drugs, memory enhancers, neuro enhancers, cognitive enhancers and intelligence enhancers. One survey found that 7% to 25% of college students used stimulants for a cognitive edge. See, "Towards responsible use of cognitive-enhancing drugs by the healthy," *Nature: International Weekly Journal of Science.* (Retrieved, 10/04/2012).

74 John L. Casti, "Reality Rules: II, Picturing the Worlk in Mathematics, The Frontier," (Wiley Interscience, 1992), p122.

75 I.Vidal-Gonzalez, et al., "Microstimulation reveals opposing influences of prelimbic and infralimbic cortex on the expression of conditioned fear," *Learning & Memory*, 13, 728-733 (2006).

76 University of Wisconsin-Madison (June 25, 2008). "How Ritalin Works In Brain To Boost Cognition, Focus Attention." *Science Daily*. (Retrieved October 7, 2012) http://www.sciencedaily.com /releases/2008/06/080624115956.htm.

77 G. Brindley, W. Lewin, "The sensations produced by electrical stimulation of the visual cortex." *J Physiol* (London) 196, 479–493 (1968).

78 E M. Maynard "Visual prostheses." *Annual Review of Biomedical Engineering*; 3:145-68. (2001).

79 W. H. Dobelle "Artificial vision for the blind by connecting a television camera to the visual cortex." *ASAIO Journal*; 46 (1):3-9 (2000).

80 Sheila Nirenberg, Chethan Pandarinath, "Retinal prosthetic strategy with the capacity to restore normal vision" http://www.pnas.org/content/early/2012/08/08/1207035109.abstract? sid=9ccf31e7-e886-45ca-bb4d-5ce813ef9b66, (Last visited, 8/23/2012).

81 Famously, in 1755 a blind man saw flames passing rapidly downwards when a voltage in a Leyden jar was discharged through the man's eye.

82 J. Pine "A History of MEA Development." M. Baudry, M. Taketani, eds. *Advances in Network Electrophysiology Using Multi-Electrode Arrays*. (New York: Springer Press, 2006).

83 R. Bhandari, et al.,"Wafer Scale Fabrication of Penetrating Neural Electrode Arrays," *Biomedical Microdevices*, Vol. 12(5), pp. 797-807, (2010).

84 Richard A. Normann, et al., "A neural interface for a cortical vision prosthesis," *Vision Research* 39 (1999) pp. 2577–2587.

85 http://www.imidevices.com/en/retinat-implantat-technology.html (Last visited 9/30/2012).

86 http://www.medtechinsight.com/advertorials/learningretina.pdf. (Last visited 9/30/2012).

87 "Stem cells may allow blind to see." *Review of Optometry*, (July 15 2012).: Also see, Motosugu Eiraku et al, "Self-Organizing Optic-Cup Morphogeneisi in Three-Dimensional Culture," *Nature* 472, pp 51–56

(April 7, 2011).

88 "Grow Your Own Eye." *Scientific American,* Vol. 307, No. 5, (Nov. 2012).

89 U.S. Pat. 8,257,272, Yang, et al. issued September 4, 2012.

90 http://online.wsj.com/article/SB100014240527023034047045773114218
88663472.html (Last visited, 8/06,2012).

91 http://www.ihealthbeat.org/articles/2012/8/3/fda-approves-ingestible-
digital-pill-to-monitor-medication-adherence.aspx (Last visited 8/4/2012).

92 John Donoghue a Brown University neuroscientist and his research team
implanted a chip into a rhesus macaque monkey's motor cortex,
establishing "a brain/machine interface" that allowed the animal to control a
computer cursor using nothing more than its thoughts. *Nature* (March 14,
2002).

93 Leigh R. Hochberg et al, "Neuronal ensemble control of prosthetic
devices by a human with tetraplegia," *Nature* 442 (2006).

94 Miguel A.L. Nicolelis, "Mind In Motion." *Scientific American,* pp. 60-
63, vol. 307, No. 3. (Sept. 2012).

95 Will Oremus, "Will the 2014 World Cup's First Kick Come From a
Mind-Controlled Robotic Leg?" *Slate* (Aug. 31, 2012),
http://www.slate.com/blogs/future_tense/2012/08/31/mind_controlled
_artificial_limbs_ could_a_robotic _leg_
make_the_world_cup_s_first_kick_.html.

96 Alan M. Turing, "Computing Machinery and Intelligence," *Mind,* 49: pp
433-460 (1950).

97 M. Mitchell Waldrop, "Computer Modeling: Brain in a Box," *Nature
News,* (Feb. 22, 2012), http://www.nature.com /news/ computer-modelling-
brain-in-a-box-1.10066.

98 Michio Kaku, *Physics of the Future,* (Anchor Books, 2012).

99 David Hubel and Torsten Wiesel received the Nobel Prize in 1981 for
their discovery that nerve cells in the visual cortex of an anesthetized cat
showed a series of responses when light was shone on the cat's eye.

100 Francis Crick, *The Astonishing Hypothesis*, (Scribner, 1994).

101 Terry Sejnowski, Tobi Delbruck, "The Language of the Brain," *Scientific American*, Vol. 307, no. 4, (Oct. 2012).

102 Birks E, Tansley, et al "Left Ventricular Assist Device and Drug Therapy for the Reversal of Heart Failure". *New Eng. J. Med* 355 (18) (2006).

103 "Successful Implantation of a Continuous-Flow, Total Artificial Heart in a Patient at The Texas Heart Institute," http://texasheart.org/AboutUs/News/2011-03-23news_tah.cfm? &RenderForPrint=1 (Last visited 9/30/2012).

104 Basic background information about molecular biology and genetic engineering, can be found in Alberts, Bray, Lewis, Raff, Roberts & Watson, *The Molecular Biology of the Cell*, 1-253, 385-481 (1983); Watson, Hopkins, Roberts, Steitz & Weiner, *The Molecular Biology of the Gene*, Vol. 1 (4th ed., 1987) 3-502.

105 G.Mendel, Versuche über Pflanzen-Hybriden. Verh. Naturforsch. (1866)Ver. Brünn 4: 3–47 (in English in 1901, J. R. Hortic. Soc. 26: 1–32).

106 Thomas Hunt Morgan, (Yale University Press,1928). On-line Electronic Edition: http://www.esp.org/books/ morgan /theory/facsimile/ (last visited, 7/25/2012.

107 The human genome contains 3,164.7 million nucleotide bases composed of (A, C, T, and G).

108 See, snp.cshl.org.

109 M Mandelkern, et al.,"The dimensions of DNA in solution," *J Mol Biol* 152 (1) (1981).

110 S Gregory, et al.,"The DNA sequence and biological annotation of human chromosome 1," *Nature* 441 (2006).

111 Discovered by Buckminster Fuller, a buckyball is a form of carbon having a large molecule consisting of an empty cage of sixty or more carbon atoms.

112 Robert A. Freitas, Jr. "Mind Uploading" (a collection of papers on the topic), available at https://www.foresight.org/Nanomedicine/Uploading.html (Last visited 10/04/2012).

113 Richard E. Smalley. "Of Chemistry, Love and Nanobots"; *Scientific American*; (September 2001).

114 Kaku, *Physics of the Future.*

115 "New method to identify intermediates in protein folding, Advancing nanotechnology with protein building blocks" *Foresight Institute at Space Frontier Conference*, Foresight Institute, http://www.foresight.org/nanodot/?p=5196 (last visited 10/01/2012).

116 J Huxley. *Evolution: The Modern Synthesis* (1942) (MIT Press, 2010). Note: Julian Huxley was the brother of Aldous Huxley, author of Brave New World.

117 J.Huxley, *Evolution in action.* (Chatto & Windus, London, 1953). p132.

118 Jean-François Lyotard, *The Postmodern Condition* (1979) (Manchester University Press, 1984).

119 Karl von Frisch discovered that knowledge about feeding places can be relayed from bee to bee through two forms of a coded message dance. The "round dance" provides information of food sources between the hive and a distance between 50 and 100 meters, and the waggle dance information about the direction and distance of food sources from the hive. See, J. Riley,et al.,"The flight paths of honeybees recruited by the waggle dance,"*Nature* 435 (2005).

120 U.S. Patent 1,647, Samuel F. B. Morse, *American Electro-Magnetic Telegraph,* 1840.

121 In 1794 Claude Chappe used a semaphore, a flag-based alphabet that depended on a line of sight for communication. In 1828, Harrison Dyar sent electrical sparks through chemically treated paper tape to burn dots and dashes. In 1825, William Sturgeon invented the electromagnet and in 1830, Joseph Henry demonstrated Sturgeon's electromagnet for long distance communication by sending an electronic current over one mile of wire activating an electromagnet.

122 Herman Hollerith to conceive the idea of using perforated cards in a system that passed punched cards over electrical contacts in a device to gather statistics during a national census. In 1889 He was granted patent protection related to the automation of tabulating and compiling statistical information. A company utilizing the Hollerith idea formed from several smaller companies in 1911, as Computing-Tabulating-Recording Company and in 1924 Thomas J. Watson, Sr. changed the company name to International Business Machine Corporation (IBM).

123 In 1946 John Eckert, Jr. and John Mauchly, (using certain ideas that Vincent Atanasoff, an engineer from Iowa shared with these gentleman), created ENIAC, short for the Electronic Numerical Integrator And Computer with hard wired program, but soon to follow the combination of the vacuum tube and magnetic core technology would permit the storing and alteration of memory data.

124 A. M. Turing, "Can a Machine Think", *The World of Mathematics*, James R. Neuman, ed., (Simon and Schuster 1956).

125 Joab Jackson, "IBM Watson Vanquishes Human Jeopardy Foes," *PC World*, (Feb. 17, 2011) Retrieved 2012-09-02.

126 H. E Kubitschek *"Cell volume increase in Escherichia coli after shifts to richer media," J. Bacteriol.* 172 (1), (January 1990).

127 "Parallel universes, the Matrix, and superintelligence" Michio Kaku, interviewed by *KurzweilAI.net* Editor Amara D. Angelica, (June 26, 2003), *http://www.kurzweilai.net/parallel-universes-the-matrix-and-superintelligence* (Last visited 7/ 10/2012).

128 The mathematician Leonard Euler posed a similar problem (The Seven Bridges of Konigsberg)in which there is an island with two branches of a river flowing on each side respectively. Seven bridges cross the two branches. The object is to plan a walk such that you cross each of seven bridges just once. These problems take the form of a situation-space or state-space where an operator or computational rule is applied to existing states to produce new states.

129 M. Kahan,et al., "Towards molecular computers that operate in a biological environment," *Physica D: Nonlinear Phenomena* 237 (9) (2008).

130 P. Frisco, et al., "Simulating Turing machines by extended mH systems," *Computing with Bio-Molecules. Theory and Experiments*, (1998), pp 221-238. Also see, P. Frisco. "A direct construction of a universal

extended H system," *Machine, Computations and Universality*, Third International conference, (2001), http://www.wi.leidenuniv.nl/home/pier.

131 This is referred to as von Neumann's Universal Constructor.

132 John L. Casti, *Reality Rules, Picturing the World in Mathematics, The Fundamentals*, (John Wiley & Sons, 1992). p211.

133 J. von Neumann, "The General and Logical Theory of Automata," *Collected works*, ed. A.H. Taub, vol 5, pp. 288-328. (New York:Pergamon,1956).

134 J. von Neuman, *Theory of Self-Reproducing Automata*, (University of Ill. Press, Urbana, IL, 1966).

135 "Self-replication of information-bearing nanoscale patterns," *Nature,* http://www.nature.com/nature/journal/v478/n7368/full/nature10500.html. Retrieved 2012-09-14. Also see, "Self-Replication Process Holds Promise for Production of New Materials," *Science Daily*, http://www.sciencedaily.com/releases/2011/10/111012132651.htm. Retrieved 2012-09-14.

136 Marcel J.E. Golay, "Reflections of a Communications Engineer,"*Analytical Chemistry,* Vol. 33, No. 7, (June 1961).

137 Casti, *Reality Rules, Picturing the World in Mathematics, The Fundamentals*.

138 John Horton Conway described this type of automata. See, "Life" *Scientific American,* (October 1970). Also see, J. Gardner, et al., *Life and Other Mathematical Amusements*, (Freeman, San Francisco, 1983).

139 M. Golay, J. Carvalko, and K. Preston investigated initial configurations that produced a spontaneous cancerous growth utilizing an hexagonal matrix of points. *Research Image*, Perkin-Elmer, Inc. Vol. 6, No. 1, (Jan. 1971).

140 Langton, et al, *Artificial Life-II*, C., eds. (Addison-Wesley, Redwood City CA, 1992); Also see, Richard Dawkins, *The Blind Watchmaker*, (Longman, London, 1986).

141 E. Regis, *Who's Got Einstein's Office* (Addison-Wesley, 1987). Also see, http://www.stephenwolfram.com/ about-sw/interviews/87-einstein/text.html.

142 Casti, *Reality Rules, Picturing the World in Mathematics, The Fundamentals.*

143 Desmond S.T. Nicholl, *An Introduction to Genetic Engineering,* (Cambridge Univ. Press, 1994).

144 Neil P. King, et al, "Computational Design of Self-Assembling Protein Nanomaterials with Atomic Level Accuracy," *Science* : Vol. 336 no. 6085 (June 2012) pp. 1171-1174. Also, see, "New method to identify intermediates in protein folding, Advancing nanotechnology with protein building blocks" *Foresight Institute at Space Frontier Conference,* http://www.foresight.org/nanodot/?p=5196 (Last visited 10/01/2012).

145 T.S. Gardner, et al., "Construction of a genetic toggle switch in Escherichia coli," *Nature* 403 (6767) (January 2000.

146 A. Levskaya, "Synthetic biology: engineering Escherichia coli to see light," Nature 438 (7067) (2005).

147 Michael Specter, "A Life of Its Own, Where will synthetic biology lead us?"*New Yorker, Annals of Science,* (September 28, 2009). http://www.newyorker.com/reporting/2009/09/28/090928fa_fact_specter?printable=true#ixzz26xAwyxUh (Last visited 09/19/2012).

148 At ETH Zurich they have infused hamster cells with networks of genes where by adding certain antibiotics turned the output of the synthetic genes to low, medium or high, which could eventually lead to gene therapies, synthesizing drugs and the manufacture of proteins.

149 U.S. Pat. No. 6,670,154 describes methods for converting modified bacterial genomes into artificial yeast chromosomes by fusing the bacteria with yeast that linearize the modified genomes. U.S. Patent Application Publication No. 2005/0019924 describes nucleic acids and methods for introducing prokaryotic genomes into eukaryotic cells as circular molecules and conversion into artificial chromosomes. WO 02/057437 describes YAC vectors containing cytomegalovirus (CMV) genomes. U.S. Pat. No. 7,083,971 describes a recombinatorial approach and system for cloning, manipulating, and delivering large nucleic acid segments. U.S. Patent Application Publication No. 2005/0003511 and Bradshaw et al., Nucleic Acids Research, 23, 4850-56 (1995) describe yeast-bacterial shuttle vectors for cloning large regions of DNA by homologous recombination. U.S. Pat. No. Appl. 20110053272; Gwynedd A. et al. March 3, 2011.

150 W. Wayte Gibbs "Synthetic Life," *Scientific American,* (May 2004).

151 Robert Lee Hotz "Scientists Create First Synthetic Cell," *The Wall Street Journal* (May 21, 2010). http://online.wsj.com/article/SB10001424052748703559004575256470152 341984.html; Retrieved Sept. 19, 2012.

152 http://www.fda.gov/AboutFDA/CentersOffices/OfficeofMedicalProduct sandTobacco/CDRH/CDRHOffices/ ucm300015.htm.

153 http://www.fda.gov/CombinationProducts/AboutCombinationProducts/ ucm101464.htm

154 *Rylands V. Fletcher,* 3 HL 330 (1868).

155 "Health Care Costs: A Primer, key information on health care costs and their impact," *Henry J. Kaiser Family Foundation,* (May 2012), http://www.kff.org/insurance/7670.cfm.

156 The question presented is analogous to the one: whom should own organs needed for transplant? According to the United Network for Organ Sharing (UNOS), nearly 116,849 Americans were waiting for an organ transplant, as of December 18, 2012.

157 There is no need to cite the imperfect, but nonetheless extensive safety net comprises charity and government assistance.

158 The public debate, legislation, regulation and litigation over the patentability of software lasted over thirty years.

159 D. Solteret al., "Putting stem cell cells to work," *Science* 282:1468 (1999).

160 See, *The Human Embryonic Stem Cell Debate, Science, Ethics, and Public Policy,* eds. S. Holland, K. Lebacqz, and Laurie Zoloth), (MIT Press, 2001).

161 DNA vaccines and other medicines for HIV, influenza are now in clinical trials.

162 *The National Organ Transplant Act (NOTA),* makes it illegal to buy or sell organs for profit. It carries a maximum sentence of five years in prison and/or a $50,000 fine. Also see, World Health Organization, *Guiding*

Principle 5, Sixty-Second World Health Assembly a62/15 Provisional
Agenda item 12.10 26 March 2009.

163 *Brenner v. Manson*, 383 U.S. 519 (1966).

164 Kurzweil writes: "We are now in a position to speed up the learning
process by a factor of thousands or millions by migrating from biological to
non-biological intelligence. *How to Create a Mind*, Ray Kurzweil, (Viking,
2012).

165 Tom Valeo, *"Where in the Brain is Intelligence?," Dana* (April 4,
2008), http://www.dana.org/news/features/ detail.aspx?id=11918 (Last
visited 10/07/2012).

166 John Rawls, wrote that "…there is no social world without loss: that is,
no special world that does not exclude some ways of life that realize in
special ways certain fundamental values", See, J. Rawls,*Political
Liberalism*, (Columbia University Press, 1993). p. 193.

167 Mark Schoofs, "Glaxo Attempts to Block Access To Generic AIDS
Drugs in Ghana," *Wall Street Journal*, (December 1, 2000).
.

168 William W. Fisher, III, et al., "The South Africa AIDS Controversy A
Case Study in Patent Law and Policy," *The Law and Business of Patents*,
(Harvard Law School, Feb. 10, 2005).

169 The price of health without frontiers Merck's decision to supply Aids
drugs to the developing world at cost could unleash a global price war, says
David Pilling, *Financial Times.com*, (March 8 2001).

170 Several lawsuits ensued against Bayer and others charging violations of
state and/or federal antitrust laws. The cases were filed in several states and
consolidated in Brooklyn, N.Y., federal court. *In re Ciprofloxacin
Hydrochloride Antitrust Litigation*, No. MDL 001383 (E.D.N.Y.).

171 Federal law, 28 USC Section 1498(a) states, in part: "Whenever an
invention described in and covered by a patent of the United States is used
or manufactured by or for the United States without license of the owner
thereof or lawful right to use or manufacture the same, the owner's remedy
shall be by action against the United States in the US Court of Federal
Claims for the recovery of his reasonable and entire compensation for such
use and manufacture." This section has been construed by the Court of
Appeals for the Federal Circuit to authorize a taking under the Fifth

Amendment's doctrine of eminent domain.

172 "Administration Won't Allow Generic Versions of Cipro," *New York Times* (October 18, 2001).

173 See, http://www.wto.org/english/tratop_e/trips_e/public_health_faq_e.htm (Last visted 12/17/2012).

174 For a thorough analysis of current patent holder success rates see, http://www.pwc.com/en_US/us/forensic-services/publications/assets/2012-patent-litigation-study.pdf.

175 Lyotard writes: "That scientific and technical knowledge is cumulative is never questioned. At most, what is debated is the form that accumulation takes - some picture it as regular, continuous, and unanimous, others as periodic, discontinuous, and conflictual." See, *The Postmodern Condition*.

176 http://bioethics.gov/cms/sites/default/files/PCSBI-Synthetic-Biology-Report-12.16.10.pdf (Last visited 8/16/2011).

177 Computer software patents have evolved such that protection runs to a solution such as a business model, such as Amazon.com or a Google search engine, that is largely a solution to one more mathematical problem. The computer itself, as a vehicle within which the software operates, has been relegated to the status of some vestigial artifact. It no longer serves to provide novelty, but merely serves as a foundational element such as an off-the-shelf engine in an automobile, where for example the novelty may be in the display console or the way the GPS allows a driver to access commercial services along a route.

178 See generally, U.S. patents 5,552,281; 5,654,173; 5,817,479; 5,969,125.

179 I refer here to logic gates embodied in transistor networks, such as AND, OR, EXCLUSIVE OR, and their primitive inversions NAND, NOR, etc. which are used to achieve the transparent operations performed when we use a computer.

180 Itakura et al., pioneered production of a gene that was not from a human source, but instead was entirely synthesized in the laboratory. See, Itakura et al., "Expression in Escherichia coli of a chemically synthesized gene for the hormone somatostatin," *Science* 198,1056 (1977).

181 Nick Bostrom, "A history of transhumanist thought", *Journal of Evolution and Technology* (2005), http://www.nickbostrom.com/papers/history.pdf. Retrieved 2012-09-02.

182 Nick Bostrom, *Transhumanist FAQ (version 2.1)* (2003), www.transhumanism.org/resources/FAQv21.pdf (Retrieved 2012-09-04).

183 Neil Postman, *Technopoly, The Surrender of Culture to Technology*, (Alfred A. Knopf, 1992).

184 A phrase employed by Blackstone in describing *"The Meme Machine"*. See, Susan Blackmore: "Memes and 'temes'" *TED2008*, http://www.ted.com/talks/susan_blackmore_on_memes_and_temes.html. (Last visited, 10/26/12).

185 John Searle suggested this particular analogy in *Mind, Language and Society*, (Basic Books, 1998).

186 "We need a name for the new replicator, a noun that conveys the idea of a unit of cultural transmission, or a unit of imitation. 'Mimeme' comes from a suitable Greek root, but I want a monosyllable that sounds a bit like 'gene'. I hope my classicist friends will forgive me if I abbreviate mimeme to meme. If it is any consolation, it could alternatively be thought of as being related to 'memory', or to the French word même..." Richard Dawkins, *The Selfish Gene* (Oxford University Press, 1989). p. 192.

187 An infomorph is a body of information that can possess emergent psychological features such as autonomy and personality.

188 "Fundamentals of Whole Brain Emulation: State, Transition and Update Representations," *International Journal of Machine Consciousness*, Vol. 4, No. 1 (2012).

189 Richard Dawkins, *River Out of Eden, A Darwinian View of Life*, p. 96, (Basis Books, 1995).

190 Dawkins, *River Out of Eden*, p. 104.

191 Dawkins, *River Out of Eden*, p. 105.

192 Iain Thomson, "Heidegger's Aesthetics," *The Stanford Encyclopedia of Philosophy* (Summer 2011 Edition), Edward N. Zalta ed, http://plato.stanford.edu/archives/sum2011/entries/heidegger-aesthetics.

193 Thomson, Iain, "*Heidegger's Aesthetics.*"

194 From the noun, teleonomy, meaning the principle that the body's structures and functions serve an overall purpose, as in assuring the survival of the organism. http://dictionary.reference.com/browse/teleonomy (Last visited 10/30/2012).

195 Huxley, *Evolution: the modern synthesis*, p576.

196 Jeremy Rifkin, *Algeny: A New Word--A New World*, (Viking 1983), writes about transhumanist goals for genetically modifying human embryos in order to create "designer babies". Also see, Leon R. Kass, "Babies By Means of In Vitro Fertilization: Unethical Experiments on the Unborn?," *N. ENGL. J. MED*. 285,1174, 1175 (1971).

197 Ervin Laszlo, *The Systems View of the World*, (George Braziller 1972). p 42.

198 W.S. Sutton, "The Chromosomes in Heredity," *Biol. Bull*. 4: 231-251. (1903). *Partial reproduction in: Classic Papers in Genetics,* (J.A. Peters, ed . (Prentice-Hall, Englewood Cliffs 1959), pp. 27-41.Also, see, L.A Martins, "Did Sutton and Boveri propose the so-called Sutton-Boveri chromosome hypothesis?" *Genet. Mol. Biol.* (1999). Vol.22, n.2, pp. 261-272, Retrieved 2011-03-03.

199 Sutton is most often credited with observing that chromosomes occur as pairs, and that egg and sperm cells receive only one set of chromosomes from each pair during the dividing or meiosis stage.

200 Venter: *On the verge of creating synthetic life.*

201 I am referring to Aristotle's eudaimonia as a final cause or telos, the ultimate purpose or good that we seek.

202 Aristotle, *Nichomachean Ethics*, Book, I.

203 Aristotle, *Nichomachean Ethics*, Book, I. the inner form, essence or meaning.

204 W. Ross Ashby describes the concept of "emergence" through the following examples:
"Ammonia is a gas, and so is hydrogen chloride. When the two gases are mixed, result is a solid-a property not possessed by either reactant... The twenty (or so) amino-acids in a bacterium have none of them the property of

being 'self-reproducing,' yet the whole, with some other substances, has this property. See, W. Ross Ashby, *An introduction to cybernetics*, (University paperbacks, 1956). p. 110.

205 Gregory Stock, *Redesigning Humans: Our Inevitable Genetic Future*, (Houghton Mifflin 2002), suggests this startling possibility.

206 Artificial chromosomes that will add genes in excess of a species normal complement will someday be "programmed" to express themselves dependent on one or another stimulus, See, Gregory Stock and John Campbell, eds., *Engineering the Human Germline: An Exploration of the Science and Ethics of Altering the Genes We pass to Our Children* (New York: Oxford University Press, 2000).

207 Perhaps the most salient example of a change in form might be realized with the addition to the current 46 chromosome complement additional chromosomes, for whatever purpose might be desired.

208 Laszlo, *The Systems View of the World*, p. 47.

209 Peter A. Corning, "The Re-Emergence of 'Emergence': A Venerable Concept in Search of a Theory," *Complexity* 7 (6): 18–30, (2002), http://www.complexsystems.org/publications/pdf/emergence3.pdf (Last visited, 9/25/2012).

210 James Hughes, *Citizen Cyborg: Why Democratic Societies Must Respond to the Redesigned Human of the Future*. (Westview Press, 2004).

211 DeBakey was the first to use an external heart pump successfully in a patient – a left ventricular bypass pump.

212 From "Dystopia: malposition of an anatomical part," *Mirriam Webster*, http://www.merriam-webster.com/medical/dystopia(Last visited 8/8/2012).

213 James Ogilvy, "Human Enhancement and the Computational Metaphor," *Journal of Evolution and Technology* - Vol. 22 Issue 1– (December 2011).

214 "Beyond Therapy: Biotechnology and the Pursuit of Happiness," *The President's Council on Bioethics,* Chapter Four, Washington, D.C. (October 2003).

215 "Vernor Vinge, Is Optimistic About the Collapse of Civilization," Wired magazine, (March 2012),

http://www.wired.com/underwire/2012/03/vernor-vinge-geeks-guide-galaxy/ (Last visited, 11/11/2012).

216 Small changes in rules or starting conditions in cellular automata result in unexpected changes in the evolution of games, population growths and life forms. It begs the question, whether any reorganization of biological constituents does not at least require us to consider where a rule change or starting condition takes us.

217 *Journal of Medicine and Philosophy*, 30:261–283, (2005) p. 265.

218 C.S. Lewis, *The Abolition of Man*, Chapter 3, (1943).

219 Alasdair MacIntyre, *After Virtue*, (Univ. Notre Dame Press, 1984). p. 216.

220 Margaret A. Farley, "Roman Catholic Views on Research Involving Human Embryonic Stem Cells," *The Human Embryonic Stem Cell Debate, Science, Ethics, and Public Policy*.

221 Lewis, *The Abolition of Man*.

222 Fukuyama, *Our Posthuman Future: Consequences of the Biotechnology Revolution*, p. 130.

223 Performing chromosomal modifications having permanent biological consequences may violate widely observed principles governing research on human subjects (see the 1964 *Declaration of Helsinki*, as amended). However, germline modifications at the preembryo stage may fall outside proscriptions against genetic manipulation of humans. See, Stuart A. Newman, "Averting the clone age: prospects and perils of human developmental manipulation," *J. Contemp. Health Law & Policy* 19: 431 (2003).

224 See, George L. Annas et al., "Protecting the Endangered Human: Toward an International Treaty Prohibiting Cloning and Inheritable Alterations," *28 AM. J. LAW MED.* 145, 151 (2002).

225 Jim Harrison, *Songs of Unreason*. (Copper Canyon Press, 2011). Reprinted with permission.

226 By 2012 estimates, 61 million children around the world do not attend school.

227 Margaret R. McLean writes: "Stem cell technology places control of the biologic processes of aging and disease as well as germ line genetics into the invisible hand of the market and the fleshy hands of individuals, making public policy formation much more complex than for older technologies that require systematic societal involvement." See, Margaret R. McClean, "Shaping the Future in Public Policy," *The Human Embryonic Stem Cell Debate, Science, Ethics, and Public Policy.*

228 In 1999, Geron Corporation Ethics Advisory Board held "hearings" on Research with human embryonic stem cells: Ethical considerations. *Hastings Center Report* 29:31-36.

229 MacIntyre argues *in After Virtue* that "...there is a fundamental contrast between man-as-he- happens-to-be and man-as-he-could-be-if-he-realized-his-essential- nature... The precepts which enjoin the various virtues and prohibit the vices instruct us how to move from potentiality to act, how to realize our true nature, and to reach our true end. To defy them will be to be frustrated and incomplete, to fail to achieve that good of rational happiness which it is peculiarly ours as a species to pursue."

www.ingramcontent.com/pod-product-compliance
Lightning Source LLC
Chambersburg PA
CBHW071548200326
41519CB00021BB/6650